```
 120
  60
 ----
 000
7 00
 ----
720 0
```

Physical Science II

Physical

Science II

Physical Science Group

Prentice-Hall, Inc., Englewood Cliffs, New Jersey

Physical Science II
Physical Science Group

© 1972 by Newton College of the Sacred Heart, Newton, Massachusetts. Copyright assigned to The Trustees of Boston University, 1974.
Published by Prentice-Hall, Inc., Englewood Cliffs, New Jersey, 07632. All rights reserved. No part of this book may be reproduced in any form or by any means without permission in writing from the publisher.
Except for the rights to material reserved by others, the publisher and the copyright owner hereby grant permission to domestic persons of the United States and Canada for use of this work in whole or in part without charge in the English language in the United States and Canada after March 1, 1977, provided that written notice is made to The Trustees of Boston University and that publications incorporating materials covered by these copyrights contain the original copyright notice, and a statement that the publication is not endorsed by the original copyright owner. For conditions of use and permission to use materials contained herein for foreign publication or publications in other than the English language, apply to the copyright owner.

Printed in the United States of America.

ISBN 0-13-671354-8 cloth
      0-13-671339-4 paper

10 9 8

Prentice-Hall International, Inc., *London*
Prentice-Hall of Australia, Pty. Ltd., *Sydney*
Prentice-Hall of Canada, Ltd., *Toronto*
Prentice-Hall of India Private Ltd., *New Delhi*
Prentice-Hall of Japan, Inc., *Tokyo*

Preface

This is the textbook for a second-year course in physical science. It is intended for students who have had the *Introductory Physical Science* (IPS) course, preferably in the preceding year.

Why two years of physical science? More specifically, why follow IPS with PS II before going to another science course such as biology, earth science, or chemistry? In the IPS course, students acquire general learning skills in reasoning, working in the laboratory, analyzing experimental data, reading scientific material, and communicating their findings to others. This is accomplished by restricting the amount of subject matter and allowing the students to study it actively in depth rather than be merely exposed to it.

All the material in the IPS course is fundamental: the characteristic properties of matter, methods of separating substances, the laws of compound formation, and an introduction to the atomic model of matter. There are other topics that are basic to further studies in science. Two broad topics of utmost importance are the connection between atoms and electric charge, and the various forms and changes in energy culminating in the law of conservation of energy. These two topics are treated in this course in the spirit of IPS.

Some of the problems in the book have been marked with a dagger (†) to indicate that short answers to them are given in the back of the book to help students check on their understanding as they proceed through a chapter.

Acknowledgments

The PS II project started in the spring of 1967, and a pilot edition was tried by 25 teachers and 550 students during the following school year. After the first year a revision was made and a new preliminary edition prepared. During the next two years, the number of pilot teachers increased to 52, and the total number of students taking the course increased to about 3,000. I wish to thank the pilot teachers for the invaluable feedback they provided. On the basis of their feedback, the present edition was prepared.

Acknowledgments

The following members of the IPS Group, in addition to myself, were involved in the development of the course: Gerald L. Abegg, Judson B. Cross, John W. DeRoy, John H. Dodge, Richard A. Hoag, Chaim Korati, Stephen McKaughan, Harold A. Pratt, Charles M. Shull, Jr., and James A. Walter.

Considerable time was devoted to the project by others who joined us for one year, for the summer, or as consultants on a part-time basis throughout the year: Mitchell Bronk, Thomas Dillon, Robert Gardner, Yehuda Elkana, Bette LeFeber, George Hall, Thomas Judd, Kemp Kolb, Elisabeth C. Lincoln, Edward Shore, and Carol-Ann Tripp.

Some experiments in this course were developed in cooperation with the Education Group at the Weizmann Institute of Science in Rehovoth, Israel, during my stay there in the academic year 1966–67.

I benefited greatly from frequent consultation with Byron L. Youtz and Aaron Lemonick, and from advice in planning the course from M. Kent Wilson.

Valuable services were provided by R. Paul Larkin and George Frigulietti in the preparation of line drawings and graphs, and by Benjamin T. Richards in production. The photographs are by George Cope, Joan Hamblin, R. Paul Larkin, and Victor E. Stokes.

Most of the administrative burden of the program was carried by Geraldine Kline. Mary Beth Nelson and Katherine Solovicos assisted in the organization of feedback and in communication with the pilot schools.

Edward M. Steele and the editorial staff of the Educational Book Division of Prentice-Hall, Inc., helped us a great deal in preparing this edition.

The development of this course was supported by a grant from the National Science Foundation. This financial support is gratefully acknowledged.

Uri Haber-Schaim
January 1972

Contents

11 Electric Charge 1

Introduction 1. A Measure for the Quantity of Charge 2. Experiment: Hydrogen Cells and Light Bulbs 4. Experiment: Flow of Charge at Different Points in a Circuit 7. The Conservation of Electric Charge 9. The Effect of the Charge Meter on the Circuit 12. Charge, Current, and Time 14. Experiment: Measuring Charge with an Ammeter and a Clock 16.

12 Atoms and Electric Charge 19

The Charge Per Atom of Hydrogen and Oxygen 19. Experiment: The Electroplating of Zinc and Lead 20. The Elementary Charge 23. The Elementary Charge and the Law of Constant Proportions 25. Experiment: Two Compounds of Copper 27. A New Look at the Law of Multiple Proportions 29.

13 Cells and Charge Carriers 32

Experiment: The Daniell Cell 33. Experiment: Zinc and Copper in Different Solutions 34. Flashlight Cells 35. Unintentional Cells and Corrosion 37. The Motion of Electric Charge through a Vacuum 41. Electrons 45. Atoms and Ions 46. The Motion of Charge Through an Entire Circuit 48. The Direction of Electric Current 49.

14 Heat 53

Experiment: Heating Different Masses of Water 54. The Calorie 56. Experiment: Heating Different Substances 58. Heat Capacity; Specific Heat 59. Experiment: Heat Lost by a Substance in Cooling 61. Experiment: Heat Capacity and Specific Heat of a Solid 62.

15 Heat and Electric Charge 65

Experiment: The Heating Effect of a Flow of Charge 65. Experiment: Heat Produced as a Function of Number of Flashlight Cells 68. Experiment: The Voltmeter 69. Heat Produced as a Function of Voltage and Charge 69. Electrical Work 71. Electrical Resistance 73. Experiment: Heat Produced in a Daniell Cell 76.

16 Where Is the Heat? 84

Heat Produced by an Electric Motor 84. Experiment: The Heat Capacity of an Electric Motor 86. Experiment: A Free-running Motor 90. Experiment: A Motor Lifting an Object 91. Experiment: Where Is the "Missing Heat"? 92.

17 Potential Energy

Gravitational Potential Energy and Thermal Energy 94. Experiment: Gravitational Potential Energy as a Function of Mass 95. Experiment: Gravitational Potential Energy as a Function of Height 100. The Change in Gravitational Potential Energy over Different Paths 100. Heat Generated by a Contracting Spring 104. Elastic Potential Energy 105.

18 Atomic Potential Energy 110

The Missing Heat When Water Evaporates 110. Experiment: Heat of Condensation of Water 112. Atomic Potential Energy 115. Experiment: Heat Produced in the Decomposition of Water 117. Chemical Energy and Heats of Reaction 120.

Contents   ix

19   Kinetic Energy   123

Experiment: Heat Generated by a Rotating Wheel 123. Another Form of Energy 125. Experiment: Kinetic Energy as a Function of Speed 128. Kinetic Energy, Mass, and Speed 129. A Re-examination of Experiments with Falling Masses 133. Thermal Energy of a Gas 135. A "Disc Gas" Machine 135. Experiment: The Effusion of Different Gases 140.

20   The Conservation of Energy   150

Laws and Definitions 150. A Review of Energy Changes 151. Experiment: A Series of Energy Changes 155. Experiment: The Energy Associated with Light 156. Radiant Energy 158. The Law of Conservation of Energy 159. Nonreversible Processes: The One-Way Street 161. Experiment: A Disc Gas with Few "Molecules" 161. Nonreversible Processes and Large Numbers of Atoms 162.

21   Energy: A Global View   165

Absorption of Radiant Energy from the Sun 165. Energy Changes in a Hurricane 166. Photosynthesis 168. Efficiency 170. The Efficiency of an Automobile 170. A Large-scale, Man-made Energy Converter 172. How Long Will the Major Energy Sources Last? 174.

Answers to Problems Marked with a Dagger   177

Index   179

Electric Charge 11

Introduction 11.1

In your work in *Introductory Physical Science* (IPS) you used electricity to decompose water. Now we shall come back to this use and examine the connection between electricity and matter.

Figure 11.1 shows the apparatus used in Chapter 6 for the decomposition of water. With only one of the electrodes connected to the battery, no water decomposes. Nothing happens unless both electrodes are connected to the terminals of the battery.

Any household electrical appliance—be it a light bulb, a motor, or a television set—has two contacts which have to be plugged in to get the device to operate.

This common characteristic, that an electrical apparatus must have two wires connecting it to a source of electricity, gave rise in the eighteenth century to the idea that when an electrical device is working something is moving through it. That something is called electric charge. When you pull out a plug, turn off a switch, or disconnect a battery, the flow of electric charge stops, and with it the operation of the apparatus.

Fig. 11.1 Apparatus used for the decomposition of water in Chapter 6.

11.2 A Measure for the Quantity of Charge

The idea of a flowing electric charge is quite attractive, because it permits us to draw in our minds a mental picture which may eventually lead to a useful model. To develop the intuitive idea of a flowing electric charge into a model, we must not only study electrical processes in more detail, but must first of all find a way to measure the quantity of electric charge that flows through a bulb, a motor, or any other device. This situation resembles the one we encountered at the beginning of *Introductory Physical Science.* We felt that there was more substance in a rock than in a pebble, but we needed the balance to enable us to make a quantitative comparison. But, while we could see the rock and the pebble directly, we cannot see electric charge either at rest or in motion. We have to look for an effect produced by moving charge which can be measured quantitatively, and which we can use to define the quantity of electric charge. You have used such indirect methods many times before, probably without noticing it. For example, you cannot see temperature directly. To measure temperature, we use the fact that substances expand when heated, and we can construct various kinds of thermometers using thermal expansion. We shall use a similar method to build a charge meter.

1† Which of the following quantities did you measure directly and which did you measure indirectly?
    a) Density of a solid.
    b) Mass of metal cube in Expt. 3.3 (*IPS*).
    c) Density of a gas.
    d) Size of molecule of oleic acid.
    e) Mass of chlorine in Expt. 8.7.

2 What do you think an electric switch does?

3 Some lights used on trucks and automobiles are mounted by means of a metal bracket on the car body. These lights have only one wire, instead of two, to be connected to a switch and battery. How do you explain the operation of such a light?

11.2 A Measure for the Quantity of Charge

In the experiment on the decomposition of water (Sec. 6.2), you noted that the longer the time the electrodes are connected to the battery, the greater the volume of both gases produced. This suggests that more charge must have flowed through the apparatus when it was connected for a longer time. Thus it seems reasonable to use the quantity of either gas produced in the reaction as a measure of the quantity of electric charge that passes through the water. We shall choose the quantity of hydrogen, since we get twice as much of this gas as we do of oxygen, and this makes

A Measure for the Quantity of Charge 11.2

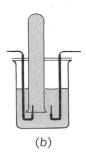

Fig. 11.2 (*a*) Apparatus for decomposing water. The electrode at the right, at which oxygen is given off, is outside the test tube as shown in (*b*), so that only hydrogen is collected.

it easier to detect small quantities of charge. This apparatus, which we shall use as a charge meter, we shall refer to as a "hydrogen cell." It is constructed as shown in Fig. 11.2. Notice that since we shall not be measuring the amount of oxygen, we have made no provision to collect it.

A source of electricity, such as a battery or a wall outlet, and one or more electrical devices connected to the source make up what is called an electric circuit. If we want to know how much charge flows through a given part of an electric circuit [Fig. 11.3(*a*) is an example], we break the circuit at that place and insert the hydrogen cell [Fig. 11.3(*b*)]. The amount of hydrogen collected tells us how much charge passed through the cell.

You will recall that volume is not a reliable measure of the quantity of matter, particularly in the case of a gas, since a gas expands and contracts appreciably as the pressure and the temperature change. Thus, if we wanted to be accurate, we should measure the quantity of electric charge in terms of the mass, rather than the volume, of hydrogen collected in the test tube. But we can be quite sure that the temperature and pressure of the hydrogen are nearly the same all over the classroom for a short

11.3 Experiment: Hydrogen Cells and Light Bulbs

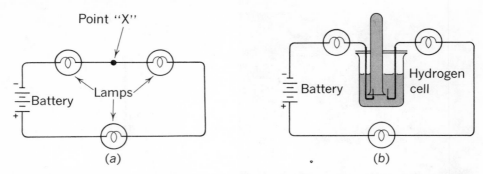

Fig. 11.3 To measure the charge that flows past point X in the circuit shown in (a), the circuit is broken, and a hydrogen cell is inserted as shown in (b).

time. Therefore, as long as we are interested only in comparing quantities of charge measured almost at the same time, we can be satisfied with simply comparing the volumes of hydrogen collected in the test tubes of different hydrogen cells. We can, therefore, choose any convenient volume of hydrogen in a test tube as our unit of electric charge. We shall use for our unit 1.0 cm³ of hydrogen.

4† An electrolytic cell for producing hydrogen and oxygen is allowed to run for 5 minutes, and then the battery terminals are reversed. It now runs for an additional 5 minutes.
  a) Does the same quantity of charge flow in each 5-minute interval?
  b) What is the ratio of the volume of gas in one tube to the volume of gas in the other?

Experiment
11.3 Hydrogen Cells and Light Bulbs

If charge flows around the electric circuit in Fig. 11.4, how will the volumes of hydrogen that will collect in the two hydrogen cells compare? Check your prediction by connecting two hydrogen cells, a battery, and a flashlight bulb as shown.

In order to collect hydrogen and not oxygen, be sure that the electrode under the test tube in each of the cells is connected to the wire which leads through the circuit to the negative (−) terminal of the battery.

Using a battery of eight flashlight cells, collect hydrogen gas until the water level has dropped about 10 cm in one of the tubes.

After disconnecting the battery, carefully mark the water level in each tube with a small rubber band or a grease pencil. You can use a graduated cylinder to measure the volume of gas collected in each test tube. Do your results agree with your prediction?

Experiment: Hydrogen Cells and Light Bulbs **11.3**

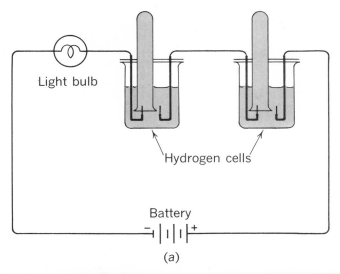

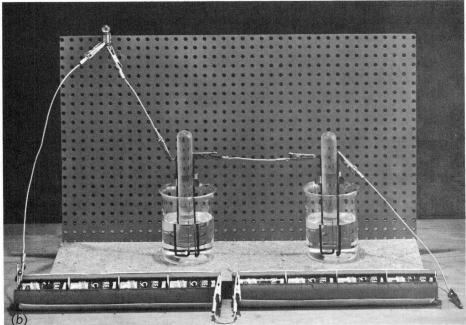

Fig. 11.4 (a) A diagram of two hydrogen cells and a light bulb connected in series to a battery of flashlight cells. (b) A photograph of the actual connections to the apparatus. The end of the battery at the left is the negative (−) terminal; the end at the right is the positive (+) terminal. Note that the negative terminal of the battery is connected, through the light bulb, to the hydrogen-producing electrode in the left-hand cell. The wire from the other electrode of this cell goes to the hydrogen-producing electrode of the right-hand cell.

11.3 Experiment: Hydrogen Cells and Light Bulbs

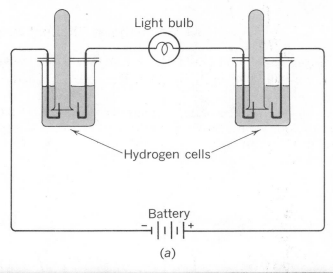

Fig. 11.5 (a) The same circuit as shown in Fig. 11.4(a), except that the light bulb is now between the two hydrogen cells so that both the charge entering the bulb and that leaving the bulb can be measured. (b) A photograph of the connections to the apparatus.

Now, rearrange the apparatus so that you can measure the charge that flows both into and out of the light bulb (Fig. 11.5). How will the gas volumes in the two hydrogen cells compare in this circuit? Do your results confirm your prediction?

Experiment
Flow of Charge at Different Points in a Circuit 11.4

The circuits shown in Figs. 11.6 and 11.7 are slightly more complex than the simple *series* circuits shown in Fig. 11.4. They are branched or *parallel*

Fig. 11.6 (a) Two hydrogen cells in parallel connected to a light bulb and a third hydrogen cell which are connected in series. (b) Apparatus connected according to the circuit in (a). The two hydrogen cells connected in parallel are on the left.

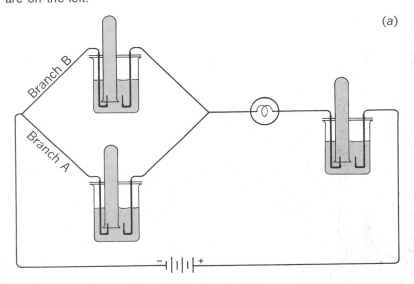

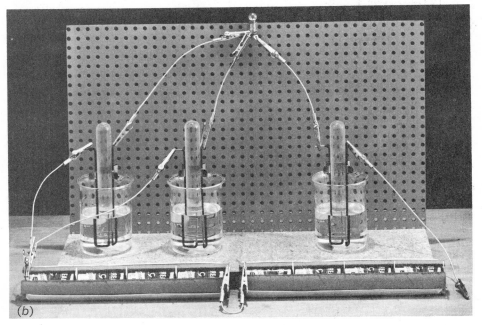

11.4 Experiment: Flow of Charge at Different Points in a Circuit

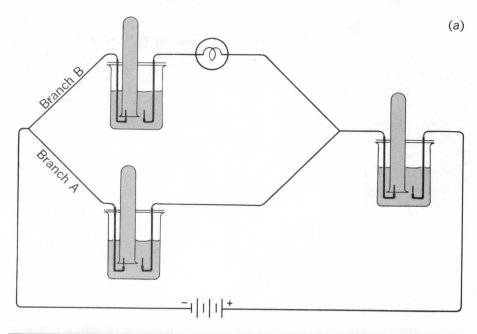

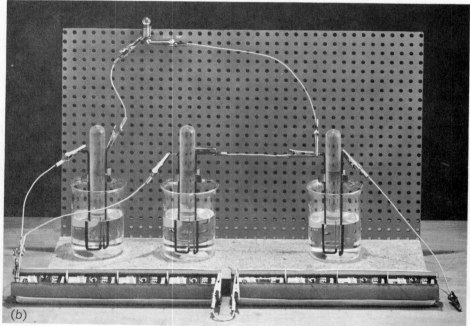

Fig. 11.7 (a) The same circuit as in Fig. 11.6(a), but with the light bulb in one of the branches of the parallel circuit. (b) The apparatus connected according to the circuit in (a). The light bulb and the hydrogen cell on the extreme left are connected in series and make up branch B of the parallel part of the circuit.

circuits with the light bulb inserted at different points. Investigate the charge flowing in each of the circuits. Again, in each case allow charge to flow until the liquid level in the test tube in the cell on the right falls about 10 cm. Can you predict what the liquid levels will be in the other tubes?

Can you find a relation between the amounts of charge flowing in different parts of the circuit?

5† A student connects to a battery a series circuit containing a hydrogen cell followed by a light bulb which is followed by another hydrogen cell. He obtains twice as much gas in one test tube as in the other. How could this be explained?

6 Each of two students connects up the circuit shown in Fig. 11.6(a) and both collect gas for the same time interval. One of the students is unaware of the fact that his connections to the battery are the reverse of those shown in Fig. 11.6(a). How would the charge he measured, in cm^3 of gas, compare with the charge measured by the student who connected the circuit correctly?

7† Two identical light bulbs are connected to a battery as shown in Fig. A. How does the charge flowing in one minute past point A compare with that flowing past points B and C?

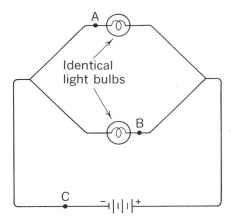

Fig. A For prob. 7.

The Conservation of Electric Charge 11.5

In the preceding section you examined the quantity of charge that passed through different parts of several electric circuits during a given time. The circuits were rather simple ones; besides the battery, hydrogen cells, and some wire, they contained only a small electric bulb. In all cases the results

11.5 The Conservation of Electric Charge

were consistent with the idea that the same amount of charge passed through all devices connected in series without being consumed.

What about a more complicated circuit—for example, one which also contains an electric motor and two radios (Fig. 11.8)? Numerous circuits containing various kinds of electrical devices have been examined, always with the same results; as long as all devices are connected in series, one after another, the results have always been consistent with the idea of conservation of charge. When two or more wires branch off from one point, that is, when parts of the circuit are connected in parallel, the sum of the charges passing through all parallel sections equals the charge that flows through the wire before the branch point. In this case as well, no electric charge is lost and none is created. Many different kinds of experiments have added support to the idea that no charge is created and no charge is destroyed.

The results of these experiments resemble those of the experiments investigating the change of mass in various processes (Sec. 2.7–2.11). Within the accuracy of the measurements, your own as well as those of others, we concluded that the total mass did not change in these reactions. We generalized these results into a law, the law of conservation of mass. The fact that within the accuracy of the measurements electric charge is neither created nor destroyed when it flows around a circuit suggests a similar generalization. It is known as the law of the conservation of electric charge: electric charge is never destroyed or created.

Just as in the case of the law of conservation of mass, we by now have so much confidence in the law of conservation of charge that when

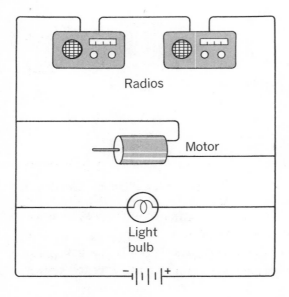

Fig. 11.8 A circuit made up of two radios in series connected in parallel with an electric motor and a light bulb.

we encounter a situation where the law seems to be violated, we check very carefully for some ways by which charge may have flowed unnoticed. For example, if a wire carrying electric charge is supported by poorly insulating materials, some charge will leak through the insulator and return to the battery through a different path. If this is not taken into account, it will look like a violation of charge conservation. But when the leaking charge is measured, conservation is found to hold.

8 Would the bulb in Fig. 11.6 have glowed with more, less, or the same brightness if it had been placed between the battery and the hydrogen cell on the right?

9† Three hydrogen cells are connected to a battery as shown in Fig. B. When 30 cm³ of hydrogen is collected in tube 1, it is observed that tube 2 has 20 cm³ of hydrogen. What is the volume of hydrogen in tube 3?

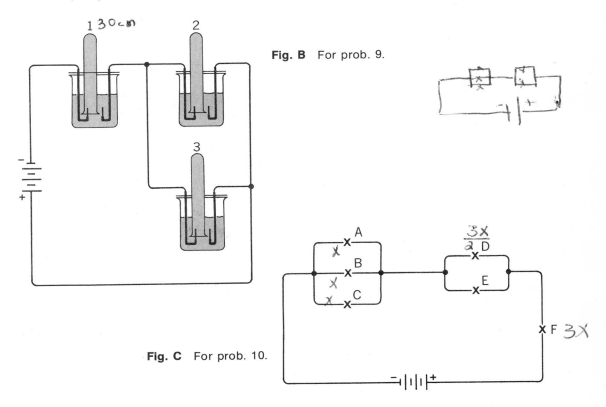

Fig. B For prob. 9.

Fig. C For prob. 10.

10 Identical hydrogen cells are inserted at the points marked "X" of the circuit shown in Fig. C. Compare the volumes of hydrogen gas that would be collected in the cells at A, D, and F in equal times.

11 In Experiment 11.4, Flow of Charge at Different Points in a Circuit, suppose that some gas escaped unnoticed while being collected in the test tube in branch A of Fig. 11.7(a). Would the results obtained suggest that charge was created or that it was destroyed?

12† In the circuits you have used so far, in a given length of time, how does the charge flowing out of the battery compare with the charge that flows into the battery?

11.6 The Effect of the Charge Meter on the Circuit

You have seen that two hydrogen cells connected in series give the same readings: a hydrogen cell does not consume charge. Does this mean that such cells do not affect the circuit at all? Figure 11.9 shows two circuits, each containing a battery and a bulb. One circuit has one hydrogen cell and the other has two hydrogen cells. Both photographs were made after the battery had been connected for 10 minutes. Notice that the water level dropped more in the circuit containing only one cell, indicating that more charge flowed in that circuit than in the other during the same time.

Thus we see that while a cell does not destroy charge, its inclusion in a circuit reduces the quantity of charge that flows through the circuit in a given time. In the example shown in Fig. 11.9, the additional cell reduced the volume of hydrogen produced in 10.0 minutes from 24.8 cm³ to 14.5 cm³. In other words, the production of hydrogen dropped from $\frac{24.8 \text{ cm}^3}{10.0 \text{ min}} = 2.48 \text{ cm}^3/\text{min}$ to $\frac{14.5 \text{ cm}^3}{10.0 \text{ min}} = 1.45 \text{ cm}^3/\text{min}$ when the second hydrogen cell was added to the circuit. We conclude, therefore, that the quantity of charge that flowed through the circuit in 1 minute with two cells was less than it was with one cell.

Such an effect on the behavior of the circuit is a very undesirable feature of the hydrogen cell as a charge meter. In general, we like any measuring instrument to have as small an effect as possible on the system to which it is applied. If this is not so, then as a result of the measurement we have a very different system than before. Suppose, for example, that to measure the air pressure in an automobile tire a pressure gauge requires so much air that after the measurement there is considerably less air, and therefore less pressure in the tire, than before the measurement was made. This would be indeed a poor gauge to use. It is worthwhile, therefore, to look for other ways to measure the quantity of charge passing through a point in an electric circuit.

A very convenient way to measure charge that eliminates the difficulties we encounter when we use a hydrogen cell is to use a clock and an instrument called an ammeter, which measures how much charge flows

The Effect of the Charge Meter on the Circuit 11.6

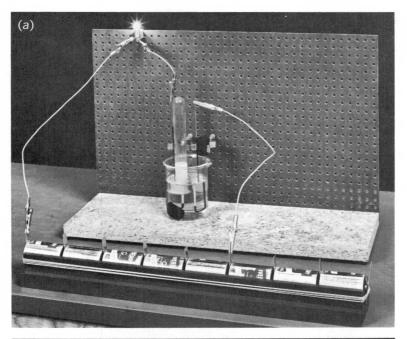

Fig. 11.9 (a) One hydrogen cell in series with a light bulb. The circuit had been operating for 10 minutes when the photograph was taken, and 24.8 cm³ of hydrogen had been produced. (b) With two hydrogen cells in series with the light bulb, only 14.5 cm³ of hydrogen was produced in 10 minutes. Note that the light bulb is much dimmer than in (a).

11.7 Charge, Current, and Time

through a circuit per unit time. Figure 11.10 shows the circuit of Fig. 11.9 with one and two ammeters instead of hydrogen cells connected in series. Notice that adding a second ammeter changed neither the brightness of the bulb nor the reading of the first ammeter. The additional ammeter had no measurable effect on the circuit.

13† Why are the masses of thermometer bulbs always much smaller than the masses of the objects whose temperatures they are made to measure?

14 Suggest an experiment to determine whether adding a light bulb to a series circuit reduces the charge that flows around the circuit in a fixed time interval.

11.7   Charge, Current, and Time

To see how a clock and an ammeter, which measures the flow of charge per unit of time, can be used to determine the total quantity of charge, first consider the following situations. A worker is paid $5 an hour or, to put it another way, he is paid at the rate of 5 dollars/hour. If he works for 8 hours, his total earnings will be 5 (dollars/hour) × 8 hours = 40 dollars. Similarly, suppose we are told that 3 gallons of water flows into a pool every second and that this flow continues for 60 sec. The amount of water that flows into the pool in 60 sec is, therefore,

$$3 \text{ (gallons/sec)} \times 60 \text{ sec} = 180 \text{ gallons.}$$

These examples suggest a general relationship:

$$\text{Total of something} = \text{(rate of something)} \times \text{(time)}$$

It is this general relationship which we shall use to measure the electric charge that passes through a point in a circuit:

$$\text{Total charge} = \text{(rate of flow of charge)} \times \text{(time).}$$

The rate of flow of electric charge, which is what an ammeter measures, is called the electric current, so,

$$\text{Charge} = \text{(current)} \times \text{(time of flow)}$$

An ammeter measures the current in units called amperes. The charge when measured by the ammeter and a clock is then expressed in units of amperes × seconds:

$$\text{Charge (amp-sec)} = \text{current (amp)} \times \text{time (sec)}$$

Charge, Current, and Time **11.7** 15

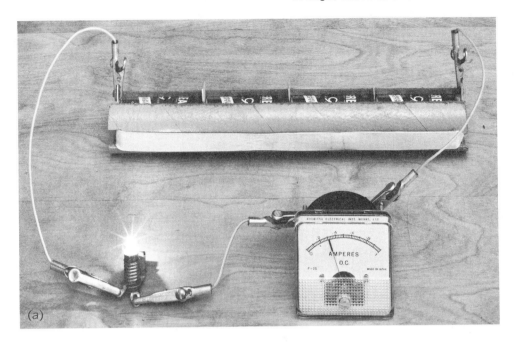

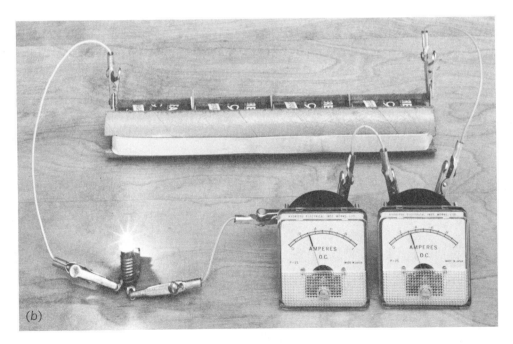

Fig. 11.10 (a) One ammeter in series with a light bulb. (b) Adding another ammeter to the circuit in (a) changes neither the brightness of the bulb nor the charge per unit time (measured in amperes on the ammeter scales) flowing through the circuit.

11.8 Experiment: Measuring Charge with an Ammeter and a Clock

If charge can be measured either with a hydrogen cell or with an ammeter and a clock, there must be a relationship between the quantity of charge measured in amp-sec and the quantity of charge measured in cm^3 of hydrogen. This relationship is the subject of the next experiment.

15† How many ampere-seconds of charge flow through an ammeter if it reads (a) 2.0 amperes for 10 seconds? (b) 0.4 amperes for 3.0 minutes? (c) 6.0 amperes for 12.0 seconds?

16† How long must current flow through an ammeter reading 0.75 amperes to have 20 ampere-seconds of charge flow through the meter?

17 You can use an ammeter and a clock to measure charge. How would you use an automobile speedometer and a clock to measure distance?

Experiment
11.8 Measuring Charge with an Ammeter and a Clock

Figure 11.11 shows a hydrogen cell and an ammeter connected in series. To find the relationship between charge measured in amp-sec and the

Fig. 11.11 An ammeter and a hydrogen cell connected in series. A graduated cylinder is used to collect the hydrogen so that the volume of gas can be measured directly. When you connect up this circuit, be sure that the positive terminal of the ammeter is connected to the positive terminal of the battery.

Experiment: Measuring Charge with an Ammeter and a Clock 11.8

charge measured in cm³ of hydrogen, you can measure the volume of hydrogen and the current at the end of every minute until the graduated cylinder is nearly full of hydrogen. Then you can use your data to make a graph of charge measured in amp-sec as a function of charge measured in cm³ of hydrogen.

Since the current may change during the experiment, you will have to know the average current. To find it, you can add all the ammeter readings up to and including the time you make a given volume reading and divide by the number of readings.

If you use between six and eight flashlight cells, you can fill the graduated cylinder with hydrogen in a reasonable time.

Use your data to draw a graph of charge expressed in amp-sec as a function of the charge expressed in cm³ of hydrogen.

Compare your graph with those of your classmates. What volume of hydrogen is produced by a flow of 1.0 amp-sec of charge? Does the volume of hydrogen produced by a flow of 1.0 amp-sec of charge depend on the number of flashlight cells used in the circuit? Does it depend on the current?

———   ———   ———

The volume of hydrogen produced by 1.0 amp-sec of charge can vary. Both the pressure and the temperature of the gas will affect the result. For example, doing the experiment in Denver, Colorado, on a hot day will give quite different results from what you would get by doing it in Boston, Massachusetts, on a cold day. However, if we calculate the *mass* of hydrogen produced in each case in 1 second, we find it is the same in both places. When one ampere flows through the cell, 1.04×10^{-5} g of hydrogen are produced per second, regardless of the temperature or pressure of the gas.

18† If the ammeter in the circuit of Fig. 11.11 measured charge instead of charge per unit time, how would the position of the needle be affected as hydrogen was produced?

19 How would the volume of hydrogen collected in Expt. 11.8 be affected by a decrease in the temperature of the collected gas during the experiment? How would this affect the value obtained for the rate of production of hydrogen gas?

20† Calculate from the data of Expt. 11.8 how many cubic centimeters of hydrogen gas at room temperature you would collect if you ran a hydrogen cell for 5 minutes with an ammeter reading of 0.5 amperes.

For Home, Desk, and Lab

21. Examine a light bulb at home and see if you can determine where the two contacts are that make the bulb operate.

22. There are no connecting wires in flashlights. Explain how it is possible for any charge to flow.

23. Suppose you have three identical light bulbs, some connecting wire, and a battery. Make a sketch of all the different possible ways to connect the battery and bulbs, using all three bulbs each time. Label which circuits are series circuits and which are parallel circuits.

24. Draw a circuit diagram including some identical hydrogen cells, showing how you could collect exactly three times as much hydrogen in one cell as in one of the others during the same time.

25. "Electric charge is neither created nor destroyed when it flows around a circuit." What happens to the electric charge when the circuit is disconnected at one point?

26. Someone suggests that the brightness of a bulb depends on the total quantity of charge that passes through it, and not on quantity of charge per unit time. How would you disprove this?

27. What would happen if the connections to the battery were interchanged (reversed) in the circuit shown in Fig. 11.11?

28. A hydrogen cell and two ammeters are connected in series. One ammeter reads 0.50 amperes and the other reads 0.65 amperes. In 10 minutes 52 cm^3 of hydrogen is produced. Using your graph from Expt. 11.8, decide which ammeter should be set aside for repair.

29. How could you use an ammeter and a hydrogen cell and the graph from Expt. 11.8 as a clock for which each cubic-centimeter mark on the 50-cm^3 graduated cylinder represents one minute?

30. How many seconds would it take to produce one gram of hydrogen in a hydrogen cell through which a current of one ampere flows?

31. Is there any evidence from the experiments you have done so far that suggests in what direction charge flows around a circuit?

Atoms and Electric Charge 12

The Charge Per Atom of Hydrogen and Oxygen 12.1

In the last chapter we used the amount of hydrogen produced by electrolysis as a measure for the quantity of electric charge that flowed in a circuit. The greater the amount of charge that passed through the cell, the greater the amount of hydrogen and oxygen produced. This indicates a connection between matter and electric charge. To investigate this connection further, we shall study electrolytic cells where the passage of electric charge results in the production of other elements on an electrode. Since we believe these elements accumulate on an electrode atom by atom, it will be useful to compare the quantities of charge needed to release single atoms of various elements.

In the case of hydrogen, a flow of 1 amp-sec of charge releases 1.04×10^{-5} g of the gas. From Sec. 9.5 and Sec. 9.7 you know that the mass of a hydrogen atom is 1.66×10^{-24} g. Thus, the number of hydrogen atoms produced by one amp-sec is equal to the mass of hydrogen produced divided by the mass of a single atom:

$$\text{Number of atoms} = \frac{\text{mass of sample}}{\text{mass of single atom}}$$

$$= \frac{1.04 \times 10^{-5} \text{ g}}{1.66 \times 10^{-24} \text{ g}} = 6.3 \times 10^{18} \text{ atoms}$$

If one amp-sec of charge releases 6.3×10^{18} atoms of hydrogen, then the charge required to release one atom of hydrogen will be

$$\frac{1.0 \text{ amp-sec}}{6.3 \times 10^{18} \text{ atoms}} = 1.6 \times 10^{-19} \text{ amp-sec}$$

12.2 Experiment: The Electroplating of Zinc and Lead

How much charge is needed to produce one atom of oxygen when water is electrolyzed? From the experiment on the decomposition of water (Sec. 6.2), you calculated that the mass ratio of hydrogen to oxygen in water is 1 to 8. From the table of atomic masses (Table 9.1), you saw that the ratio of the atomic mass of hydrogen to that of oxygen is 1 to 16. Therefore, in a compound which has one atom of hydrogen for every atom of oxygen, the mass ratio of hydrogen to oxygen would be 1 to 16. In water this mass ratio is 1 to 8 or 2 to 16. We conclude, therefore, that in water there are two hydrogen atoms for every oxygen atom. (This yields the well-known simplest formula for water: H_2O). In other words, if we decompose any amount of water, we get twice as many atoms of hydrogen as of oxygen.

Since the same quantity of electric charge passes through both electrodes, each atom of oxygen must require twice the charge needed to release one atom of hydrogen. Therefore we conclude that it takes $2.0 \times (1.6 \times 10^{-19})$ amp-sec of charge to release one atom of oxygen.

1† During the electrolysis of water, which collecting tube has the greater number of atoms of gas in it at a given time?

2 One amp-sec of charge yields 1.04×10^{-5} g of hydrogen. How many grams of oxygen will 1 amp-sec of charge release? What volume will the oxygen occupy?

Experiment
12.2 The Electroplating of Zinc and Lead

We have just seen that there is a simple, 2-to-1 ratio for the quantities of charge needed to release one atom of oxygen and one atom of hydrogen. We can release elements other than hydrogen and oxygen by electrolysis. How does the charge per atom compare for different elements?

In this experiment, we shall determine the quantity of charge needed to release one atom of zinc from a solution containing zinc, and the charge required to release one atom of lead from a solution containing lead. Since both elements are solids, they will be deposited as solids on the electrode, and we can determine the mass released by massing this electrode before and after we electrolyze the solution. The charge per atom can then be found from

$$\text{Charge/atom} = \frac{\text{charge needed to release sample of element}}{\text{number of atoms in sample}}$$

Experiment: The Electroplating of Zinc and Lead   12.2

With an ammeter and a clock we can measure the charge in ampere-seconds. From the change in mass of the electrode and the mass of a zinc atom we can determine how many atoms of zinc were plated during the experiment. The mass of a zinc atom is given in Table 12.1.

The zinc-plating cell consists of two zinc electrodes and a solution containing zinc. The lead-plating cell has two lead electrodes and a solution containing lead.

After massing each of the two zinc electrodes, you can connect the zinc-plating cell to an ammeter as shown in Fig. 12.1. Do not connect the battery to the ammeter and the plating cell until you are ready to time the run.

Start by connecting the circuit to only one flashlight cell, but quickly increase the number until you get the maximum current you can get and still have the ammeter needle on the scale.

A run of 20 to 25 minutes will deposit enough zinc to be massed on the balance. Since the current may change during the run, a current reading every minute will be useful.

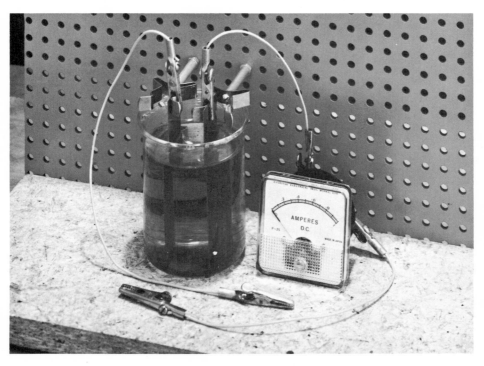

Fig. 12.1 A zinc-plating cell and an ammeter connected in series. The wire lead on the right comes from the ammeter terminal marked "+" and is connected to the positive terminal of the battery (not shown in the photograph).

12.2 Experiment: The Electroplating of Zinc and Lead

Table 12.1

Atom	Symbol	Mass in Amu	Mass in Grams
Aluminum	Al	27.0	4.5×10^{-23}
Calcium	Ca	40.1	6.7×10^{-23}
Carbon	C	12.0	2.0×10^{-23}
Chlorine	Cl	35.5	5.9×10^{-23}
Copper	Cu	63.5	1.06×10^{-22}
Gold	Au	197	3.3×10^{-22}
Helium	He	4.0	6.7×10^{-24}
Hydrogen	H	1.0	1.66×10^{-24}
Iron	Fe	55.8	9.3×10^{-23}
Lead	Pb	207	3.4×10^{-22}
Magnesium	Mg	24.3	4.0×10^{-23}
Nickel	Ni	58.7	9.8×10^{-23}
Nitrogen	N	14.0	2.3×10^{-23}
Oxygen	O	16.0	2.7×10^{-23}
Silver	Ag	108	1.80×10^{-22}
Sodium	Na	23.0	3.8×10^{-23}
Uranium	U	238	4.0×10^{-22}
Zinc	Zn	65.4	1.09×10^{-22}

When the run is over, rinse both electrodes by dipping them in a beaker of water. Gently dry the electrodes with paper towels and mass each one on your balance. Compare the change in mass of one electrode with that of the other. Use Table 12.1 to find how many atoms of zinc were deposited.

From the average current that flowed through the circuit and the duration of the run, calculate the electric charge in ampere-seconds that was used to plate out the zinc. How much charge was needed to deposit one atom of zinc? How does this compare with the charge to release one atom of hydrogen?

Now repeat this experiment, using two lead electrodes like those shown in Fig. 12.2 and a solution containing lead. How much charge is needed to deposit one atom of lead? How does the charge per atom of lead compare with the charge per atom of hydrogen?

3. A nickel-plating cell was run for 10 min at an average current of 0.80 amp. It was found that 0.15 g of nickel was deposited on one electrode.
   a) What is the mass of one atom of nickel?
   b) How many nickel atoms were deposited?
   c) What charge flowed in amp-sec?
   d) What is the charge per atom for nickel in this experiment?

4. In Expt. 12.2, The Electroplating of Zinc and Lead, one student calculated the charge in amp-sec for each minute, added all his results, and then divided

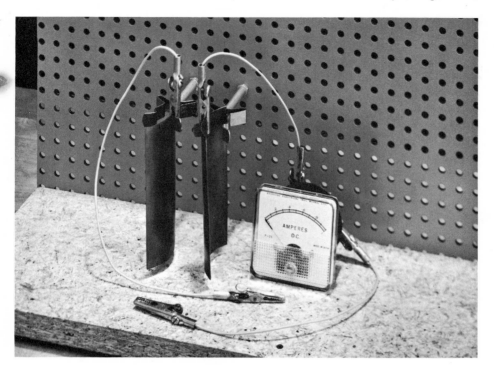

Fig. 12.2 The two lead electrodes are bent into a curved surface by pressing them against the outside of a beaker. This is to ensure an even deposit of lead that will not fall off the negative electrode. Be sure the beaker into which the electrodes are placed is filled with solution to about a centimeter from the top.

by the total time in minutes. Another student found the average current and multiplied by the total time in seconds. Which method is correct for determining the total charge?

5† A nickel-plating cell is run for 5 min at a current of 0.6 amp. Then the battery leads are reversed, and the cell runs for 5 min more at the same current. What would you expect to have happened at the electrodes? Would the result be the same for a cell electrolyzing water?

The Elementary Charge 12.3

Hydrogen, oxygen, zinc, and lead are, of course, not the only elements which can be collected at the electrodes of an electrolytic cell. Other elements can also be collected this way, and the charge needed to deposit one atom can be determined. However, the procedure may in some cases

12.3 The Elementary Charge

be considerably more difficult than the ones you have used. For example, sodium cannot be plated out from a solution of sodium chloride in water. Hydrogen will evolve at the negative electrode, and the sodium will remain in the solution. Chlorine will bubble up at the other electrode, but since chlorine is quite soluble in water, the amount of chlorine collected as gas in the test tube is much less than the amount of chlorine actually produced. To obtain reliable data from which to calculate the charge per atom for sodium and chlorine, one has to pass a current through molten sodium chloride and prevent any contact of the sodium with water vapor or oxygen. Potassium requires a similar treatment.

Despite the technical complications, many elements have been produced by electrolysis, and the charge required per atom has been determined. Table 12.2 shows the ratios of the charge per atom for some common elements to the charge per atom of hydrogen, as obtained from experiments more precise than those you have done.

Table 12.2 strongly suggests that the electric charge needed to produce one hydrogen atom by electrolysis has a fundamental significance. First of all, it is the smallest quantity of charge that is involved in the electrolysis of any material. But, more than that, the charge required to deposit one atom of any element is equal to the charge per atom of hydrogen or to a small whole number times this charge. There is no element that requires, say, 2.5 times this charge to release one atom. For these reasons the charge needed to release one hydrogen atom, $1.60 \times$

Table 12.2

Element	Symbol		Charge Per Atom of Element / Charge Per Hydrogen Atom
Aluminum	Al	4.8×10^{-19}	3.00
Bromine	Br		1.00
Calcium	Ca		2.00
Chlorine	Cl		1.00
Chromium	Cr		6.00
Iodine	I		1.00
Lithium	Li		1.00
Magnesium	Mg		2.00
Mercury	Hg		2.00
Nickel	Ni	3.2×10^{-19}	2.00
Oxygen	O		2.00
Potassium	K		1.00
Silver	Ag		1.00
Sodium	Na		1.00

(handwritten annotations: "WHEN MULT- CHARGE/ATOM"; "$\times 1.6 \times 10^{-19}$")

10^{-19} amp-sec, is known as the elementary charge. It is the smallest quantity of charge now known to occur in nature.

About 150 years ago Michael Faraday in England carried out electrolysis experiments similar to the ones you have done. He found that the mass of an element deposited on an electrode is proportional to the quantity of charge that flows through the circuit. But Faraday, who was not convinced that matter was composed of atoms, did not suggest an elementary unit of charge.

The first to propose the idea of an elementary charge was Hermann Helmholtz in Germany. In 1881 he related the results of Faraday's experiments to the atomic theory with this bold declaration: "If we accept the hypothesis that the elementary substances are composed of atoms we cannot avoid the conclusion that electricity also . . . is divided into definite elementary portions, which behave like atoms of electricity."

So far, we have seen that the units we use to measure physical quantities are quite arbitrary: the centimeter is as good a unit of length as the inch. Nowhere have we seen any preference in nature for one unit over another. The electric charge is different. Nature provides us with its own fundamental unit, the elementary charge.

6 Five opaque plastic bags are filled with varying numbers of identical marbles. The masses of the bags are as follows: 150 g; 60 g; 135 g; 195 g; 240 g. What could be the mass of a single marble (the "natural" or "elementary" unit of mass for the bags of marbles)?

7† A hydrogen cell is connected in series with a lead-plating cell.
a) How many atoms of hydrogen will be produced for every atom of lead deposited?
b) How many grams of hydrogen will be produced for every gram of lead deposited?

8 Will an electric current which passes through a sodium-plating and a lead-plating cell in series ever deposit the same mass of metal in each cell?

9 What is the significance of recording the ratios in Table 12.2 as 3.00, 1.00, 2.00 . . . , and not simply as 3, 1, 2, . . . ?

The Elementary Charge and the Law of Constant Proportions 12.4

You have seen in Chapter 9 that if you know the mass ratio in which two elements combine to form a compound, and also their atomic masses, you can find the simplest formula of the compound. For example, sodium

12.4 The Elementary Charge and the Law of Constant Proportions

and chlorine combine in the mass ratio of sodium to chlorine of 0.648 to form sodium chloride. This is exactly the ratio of their atomic masses:

$\frac{23.0 \text{ amu}}{35.5 \text{ amu}} = 0.648$. (See Table 12.1) We conclude that in sodium chloride

there is one atom of sodium for every atom of chlorine, and we assign the formula NaCl to the compound. Here is another example: calcium combines with chlorine in a mass ratio of 0.565, whereas the ratio of their

atomic masses is $\frac{40.1}{35.5} = 1.13$, just twice as much. Hence, each atom of

calcium must combine with two atoms of chlorine, and the formula for calcium chloride is $CaCl_2$.

When we break up NaCl by electrolysis, we get equal numbers of atoms of chlorine and sodium. Since the charge passing through both electrodes is the same, the charge per atom must also be the same. On the other hand, in electrolyzing $CaCl_2$ we get twice as many atoms of chlorine as of calcium. Again, since the same charge passes through both electrodes, each atom of calcium liberated must need twice the charge required for an atom of chlorine liberated, thus agreeing with the simplest formulas we obtain using atomic masses and mass ratios in these compounds.

We can now try to generalize these results as follows: Consider two elements A and B which combine to form a compound. If A and B require equal numbers of elementary charges per atom in electrolysis, then they combine according to the formula AB, that is, one atom of A for every atom of B. On the basis of this generalization, we would predict from Table 12.2 that hydrogen and iodine combine according to the formula HI, and that mercury and oxygen form the oxide HgO. When these compounds are analyzed, this is indeed found to be their composition.

If element A requires one elementary charge per atom and element B requires two, then if they combine at all they will do so according to the formula A_2B (or BA_2), since each time two elementary charges release one atom of B, two atoms of A are released. The compounds H_2O, Na_2O, and $MgCl_2$ are examples of this kind of composition.

Accurate measurements of the mass ratios of the elements in these and many other compounds confirm predictions based on Table 12.2. Additional examples are given in Table 12.3.

To sum up, once we have found the number of elementary charges per atom for various elements by electrolysis, we know how many atoms of one element would combine with one atom of another element, even

Table 12.3

Element	Elementary Charge Per Atom	Element	Elementary Charge Per Atom	Simplest Formula of Compound
Iron	2	Chlorine	1	$FeCl_2$
Silver	1	Chlorine	1	$AgCl$
Hydrogen	1	Bromine	1	HBr
Lithium	1	Oxygen	2	Li_2O
Silver	1	Oxygen	2	Ag_2O

though some compounds cannot be electrolyzed. The number of elementary charges per atom and the atomic mass fix the proportions in the law of constant proportions. When we first encountered this law, we had no way of relating mass ratios to anything else. Now we see that they are closely connected to atomic masses and the number of elementary charges per atom.

But two elements can sometimes form more than one compound. Does this mean that the same element can have different charges per atom in different compounds? We shall try to answer this question in the next experiment.

10 One can write the simplest formula for water either from the data on the combining masses and atomic masses of hydrogen and oxygen, or from the data on the charge per atom for these two elements in electrolysis. Upon what conservation laws are these arguments based?

11† Using Table 12.2, write the simplest formulas for (a) magnesium oxide, (b) chromium oxide, (c) aluminum oxide.

12† Element A combines with element B according to the simplest formula A_2B. Element B has been found to require two elementary charges per atom in electrolysis. If element A could be plated out in electrolysis, how many elementary charges per atom would it require?

13 In the synthesis of zinc chloride (Expt. 6.4) you showed that the ratio of zinc reacted to product formed was 0.48. Using the data from Tables 12.1 and 12.2, what ratio do you obtain?

Experiment
Two Compounds of Copper 12.5

The two compounds of copper to be investigated are copper sulfate (a blue solution) and copper chloride (an almost colorless solution).

12.5 Experiment: Two Compounds of Copper

Fig. 12.3 Two copper-plating cells connected in series. The wire lead on the right runs from the ammeter terminal marked "+" to the positive terminal of the battery (not shown). Note which electrodes are labeled for identification. The cell on the left contains the dark-blue solution, and the one on the right contains the almost colorless solution.

Mass two copper electrodes after marking them with identifying letters and your initials.

Connect two copper-plating cells, each with copper electrodes, in series with an ammeter as shown in Fig. 12.3. In each cell the negative electrode is the one you have marked for identification. Add the blue solution to one cell and the almost colorless solution to the other.

Using the same procedure as in Expt. 12.2, adjust the current close to 1 amp in order to deposit in 15 minutes enough copper to mass. Be sure you record the current every minute during the run.

After the marked electrodes have been carefully rinsed and dried, you can mass them and find the gain in mass of each.

How many atoms of copper were deposited in each electrolytic cell by the charge that flowed through the circuit? Without further calculation, what can you say about the quantity of charge required to plate out one atom of copper from the blue solution, compared with the charge needed to plate out one atom of copper from the colorless solution?

From the average current that flowed through the circuit and the duration of the run, calculate the number of elementary charges needed to plate out an atom of copper from each of the two solutions.

There are two oxides of copper—that is, two compounds containing only copper and oxygen. On the basis of the results of this experiment, what do you predict are their simplest formulas?

14† If in the experiment described above, the electrodes in one of the cells were bent so that they were touching throughout the run, what effect would this have on the results?

15 In Expt. 8.7 you found that twice as much chlorine combined with a given mass of copper in one chloride of copper as in another. What could be the simplest formulas for these substances, on the basis of that experiment alone? What limitations are placed on these choices by Expt. 12.5 and Table 12.2?

A New Look at the Law of Multiple Proportions 12.6

When we first introduced the atomic model of matter in Chapter 8, we made the following prediction: If two elements form more than one compound, the ratio of the masses of one element that combine with a fixed mass of the other is given by a ratio of small whole numbers. You checked this prediction in Sec. 8.7. You found that when chlorine combined with copper to form two compounds, for a fixed amount of copper the ratio of the masses of chlorine in the two compounds was 2:1. You had no way of predicting what those small whole numbers might be. Now you are in a better position. With the results of the last experiment you can tell what the numbers would be for copper compounds.

There are other elements besides copper which require different numbers of elementary charges per atom to release them, depending on the compound from which they are plated out. Table 12.4 lists some of them, including a few which have appeared already in Table 12.2.

Table 12.4

Element	Symbol	Number of El. Ch. Per Atom for Different Compounds
Chromium	Cr	2, 3, 6
Iron	Fe	2, 3
Mercury	Hg	1, 2
Tin	Sn	2, 4

For Home, Desk, and Lab

We can now use the reasoning we employed in Sec. 12.4 to predict the simplest formulas for different compounds containing these elements. Some examples are shown in Table 12.5.

Table 12.5

Element	Elementary Charge Per Atom	Element	Elementary Charge Per Atom	Simplest Formula of Compound
Iron	2	Oxygen	2	FeO
Iron	3	Oxygen	2	Fe_2O_3
Tin	2	Chlorine	1	$SnCl_2$
Tin	4	Chlorine	1	$SnCl_4$
Chromium	2	Oxygen	2	CrO
Chromium	3	Oxygen	2	Cr_2O_3
Chromium	6	Oxygen	2	CrO_3

The analysis of these and other compounds confirms the predictions. The observed number of elementary charges needed to produce one atom of an element in an electrolytic cell allows us to predict the possible compounds which that element may form when combining with any other element whose charge per atom is also known. However, the fact that you can predict from electrolysis experiments the simplest formula of a compound gives no assurance that the compound actually exists.

16† There are two compounds containing only chlorine and mercury. What would you predict as the simplest formula for these two compounds?

17 From the data in Table 12.4, write the simplest formula for the two oxides of mercury.

For Home, Desk, and Lab

18 How much charge in ampere-seconds must pass through a hydrogen cell to produce a test tube of hydrogen (about 35 cm³)? Use the class results for Expt. 11.8.

19 In the electrolysis of water we concluded that water must be continually added to the cell if the same water level is to be maintained in the cell. From your observations of metal-plating cells, do you think that the solutions involved must be replenished if the same liquid level is to be maintained?

20 A student ran a lead-plating experiment with a battery that was known to contain cells that were weak from constant use. How would this affect his calculations of charge per atom for lead?

21 In a metal-plating experiment, the following measurements and calculations were obtained.
   1. Mass of metal deposited.
   2. Number of atoms deposited.
   3. Average current.
   4. Number of charges that flowed.
   5. Charge per atom of metal.

   How would each of the above observations be affected if
   a) the length of time for the run was doubled?
   b) the electrodes were moved closer together?
   c) an additional cell was used in the battery?
   d) the electrodes were not rinsed before the final massing?

22 Three electrolytic cells were connected in series. Hydrogen was collected in a test tube, and silver and lead were plated out. If 1 g of hydrogen is collected, what would be the mass of the elements plated out?

23 Initially the unit of electric charge was the quantity of charge that released 1 cm³ of hydrogen in a hydrogen cell. Later, we found that a unit of charge called an ampere-second released 1.04×10^{-5} g of hydrogen. In this chapter, the unit of charge is associated with the release of one atom of hydrogen. Be prepared to discuss the reasons for redefining the unit of electric charge as we have done.

24 The decomposition of uranium chloride yields UCl_4 as the simplest formula for the compound. How many elementary charges per atom do you expect to find in plating uranium out of a solution?

25 Some of the first precise determinations of the atomic masses of many elements were made by the use of plating cells. Suppose a gold-plating cell and a silver-plating cell are connected in series to a battery. When 1.00 g of silver is deposited, 1.83 g of gold is plated out in the gold cell. Knowing that the mass of a silver atom is 108 amu and that one elementary charge is needed to deposit one silver atom, determine the mass of a gold atom. What important assumption did you make? How does your value for the mass of a gold atom compare with that in Table 12.1?

26 Element A combines with element B according to the simplest formula A_2B. Element B combines with element C in the simplest formula CB. What might be the simplest formula for a compound made from A and C?

27 A hydrogen-oxygen cell and a copper-plating cell containing the blue solution copper sulfate are connected in series until 10 cm³ of hydrogen is collected. Write down in words the steps which you would take to calculate how much copper was plated during this time.

13 Cells and Charge Carriers

We have drawn far-reaching conclusions about the flow of charge in a solution from our experiments with electrolytic cells. In these experiments a battery of flashlight cells was needed to cause an electric charge to flow through various solutions. What happens inside a flashlight cell when we draw current from it? It is not easy to answer that question from simple observation, but a similar process takes place inside another kind of cell, called a Daniell cell, which also produces electric current and which we can examine in detail.

A Daniell cell is constructed as shown in Fig. 13.1. Two solutions are placed in the same container, but they are separated by a thin wall of parchment paper through which they can mix only very slowly. Notice that this cell differs from the plating cells you have used, not only because two solutions are used but because the two electrodes are made of different metals. A photograph of the cell is shown in Fig. 13.2.

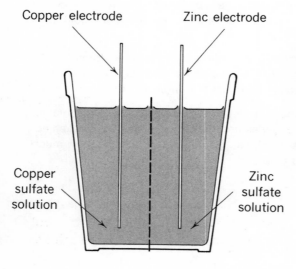

Fig. 13.1 A diagram showing the construction of a simple Daniell cell. The solutions are contained in a styrofoam cup that is divided down the middle into two compartments by a parchment paper partition shown as a dashed line in the figure.

Experiment: The Daniel Cell **13.1**

Fig. 13.2 A Daniell cell with one terminal connected to an ammeter. The cell will begin operating when the circuit is completed by clipping together the two alligator clips lying in front of the meter. The left half of the cell contains the copper electrode and the dark-blue copper sulfate solution. The right half contains the zinc electrode and the colorless zinc sulfate solution. The copper electrode is connected to the "+" terminal of the ammeter.

When a Daniell cell is providing current, is there a change in mass of the two electrodes? If so, is the charge per atom needed to dissolve or plate material the same as when zinc and copper are plated out in electrolytic cells?

Experiment
The Daniell Cell   13.1

When you have made a cell like that in Fig. 13.1, mass the electrodes and put them in place in the cell. In order to minimize mixing of the solutions, do not pour the two solutions into the cells until you are ready to record the current and time.

When the cell has been filled, connect it to the ammeter. Adjust the separation of the electrodes by rotating the clamps that hold them until the current is about 0.8 amp. It is best to record the current every minute for 15 minutes.

34   13.2   Experiment: Zinc and Copper in Different Solutions

After rinsing both electrodes, dry and mass them, and calculate the charge per atom for the mass change of each electrode. How do these values compare with those that are obtained when these metals are plated out in electrolytic cells like those you used in experiments in Chapter 12?

1. A Daniell cell will deliver current for a long time but will eventually stop. What do you think could happen to the electrodes or solutions to cause the cell to stop?

2. A Daniell-cell experiment gives 1.7 elementary charges per atom for both zinc and copper. What would be the most likely source of this error?

Experiment
13.2   Zinc and Copper in Different Solutions

The Daniell cell you have just used contained two different solutions as well as two different electrodes. You may have wondered why it was necessary to use two solutions and keep them separated by a special partition. You can answer this question by observing the effect of placing each of the electrodes used in the Daniell cell in each of the two solutions. Be sure to rinse the electrode in water each time you remove it from a solution.

First try dipping the copper electrode into some copper sulfate solution. What happens? Now try the copper electrode in zinc sulfate solution. Does anything happen?

What is the result of placing the zinc in the zinc sulfate solution? Of placing the zinc in copper sulfate solution?

In which of these cases, using just a single electrode, did you observe an effect?

Now connect two zinc electrodes to an ammeter and then dip them first into one solution and then into the other. What do you observe? Try two copper electrodes, connected by an ammeter, in each of the solutions. What happens?

Now connect a zinc electrode and a copper electrode to an ammeter and dip the two electrodes into each of the two solutions in turn.

In which of these cases, using two electrodes and an ammeter, did you observe an effect?

Suppose you have a cell consisting of electrodes of copper and zinc immersed in a copper sulfate solution. What will happen in this cell when it is not being used? Could you use such a cell to determine the number of elementary charges per atom when zinc dissolves and copper plates out?

From the observations you made in this experiment, why are the two solutions separated by parchment paper in the Daniell cell?

Flashlight Cells 13.3

The flashlight cells that you have used in your experiments, though quite complex, have much in common with a Daniell cell or a simple zinc, copper, and copper sulfate cell. Like all current-generating cells, a flashlight cell consists basically of two different electrodes and a water solution of a compound. Figure 13.3 shows the construction of a flashlight cell.

Fig. 13.3 A cross section of a flashlight cell. The porous separator keeps the carbon and manganese dioxide from coming into contact with the zinc but allows ammonium chloride solution to pass freely. The zinc can is enclosed in a tight-fitting cardboard tube which holds the insulating top in place and insulates the negative zinc can from other, adjacent cells or adjacent metal objects such as the tube of a flashlight.

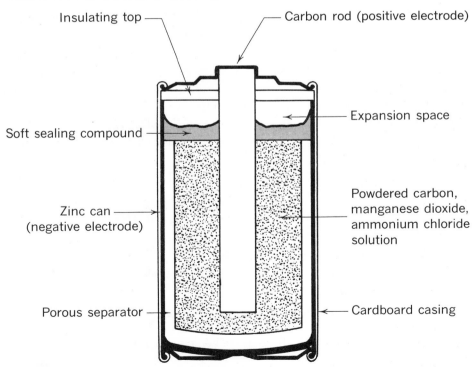

13.3 Flashlight Cells

The positive electrode is a cylindrical rod of carbon in the center of the cell, surrounded by a paste consisting primarily of a water solution of ammonium chloride mixed with powdered carbon and manganese dioxide. The negative electrode is a cylindrical zinc can that serves not only as an electrode but also as a container for the carbon electrode and the other materials. This kind of cell is often called a "dry" cell, because the solution, held in the porous paste and sealed in the zinc can, will not run out of the cell.

In this type of cell, when charge flows, zinc dissolves, forming zinc chloride; manganese dioxide then reacts with the solution to form another oxide of manganese.

If the current supplied by a dry cell is too large, hydrogen gas accumulates around the carbon electrode. As this gas accumulates, it blocks the flow of charge and the cell "runs down" rapidly to the point where it cannot supply enough current to be useful. However, such a "run down" cell partially recovers in a few hours when disconnected from a circuit as the hydrogen gas slowly leaks out of the cell. After the cell has been used for some time, insoluble, nonconducting compounds form that hinder the flow of charge through the cell. Then it cannot supply enough current to be useful and must be discarded. A newly manufactured cell, even when not used, becomes useless after a number of months or a few years, because water evaporates from the cell and reactions go on, though very slowly, in the cell even when it is not connected to a circuit.

There are many other types of cells that are used to move charge through a circuit, some of which can be "recharged." In "recharging," the reactions that go on when the cell is used are reversed by sending charge through the cell in the direction opposite to that in which the cell moves charge through a circuit. This results in the electrodes and solution being restored to their original condition, and the cell can then be used again to operate a circuit.

All cells, regardless of the materials of which they are made, are basically the same. They have two dissimilar electrodes immersed in solutions, and when they provide current, some compounds are decomposed and new ones are formed.

3   A current is drawn from a flashlight cell.
    a) How much charge in amp-sec flows if the cell delivers 1 amp for an hour?
    b) How many elementary charges is this?
    c) How many atoms of zinc dissolve?
    d) What mass of zinc dissolves? (Use Table 12.1).

4   The circuit shown in Fig. 11.11 is allowed to operate until 10 cm^3 of hydrogen is collected.

a) What is the mass of this quantity of hydrogen?
b) How much zinc dissolved in the flashlight cell at far left?
c) How much zinc dissolved in each of the other cells?

Unintentional Cells and Corrosion 13.4

It takes careful design to produce an efficient dry cell or an automobile-battery cell that will give reasonably large currents over a long time. However, it is just as hard to avoid having cells that will produce very small currents. A large variety of elements and compounds will act as electrodes, and there are many solutions which will complete the cell. For example, an aluminum nail and a copper nail stuck into an olive, a lemon, or almost any other fruit will produce a current, though a small one (Fig. 13.4). Even tap water has enough dissolved material to make a current-producing cell when two dissimilar metals are immersed in it. The sweat from your hands, or even saliva, will also serve as a solution for a cell.

Often, two dissimilar metals and a solution may accidentally form an unwanted current-producing cell. Such a cell results when a piece of iron and a piece of copper are lying on moist earth and are touching each other. The iron metal will rust away rapidly, and a sensitive ammeter would show that charge flows between the two metals where they are in contact. Unintentional cells like this are the cause of an enormous amount of corrosion damage to metal structures because of the tendency of some metals to react when they are in contact with other metals (or in some cases compounds) in a moist environment.

Fig. 13.4 On the left is a current-producing cell made from a lemon. The two electrodes are an aluminum nail and a copper nail. The solution in the cell is the juice of the lemon. On the right is a similar cell made from an aluminum nail, a copper nail, and an olive. The current produced by these cells is very small, and the scale on the ammeter in the photographs is in millionths of an ampere.

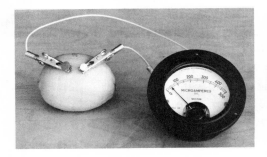

13.4 Unintentional Cells and Corrosion

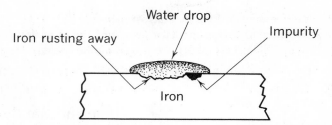

Fig. 13.5 The rusting of iron is speeded by the action of small, current-producing cells created by bits of impurity embedded in its surface. In the drawing a water drop, containing a small amount of dissolved material, serves as the "solution" of the tiny cell, and charge flows directly from the impurity into the iron.

Fig. 13.6 An iron tank buried in damp soil can be protected against corrosion by making it one electrode of a current-producing cell of which the other electrode is a plate of magnesium. The magnesium slowly dissolves, but the iron of the tank does not. In practice, several magnesium electrodes are buried around the tank, each being connected to it by a wire.

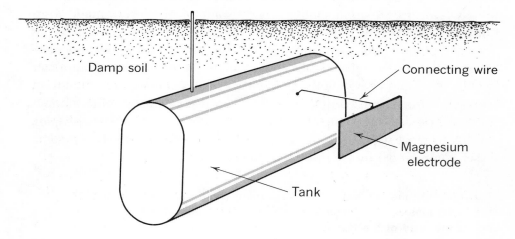

Very pure iron does not rust rapidly. But most iron and steel contains impurities such as carbon, some in the form of little islands embedded in the surface of the metal (Fig. 13.5). When the surface of the iron or steel becomes damp, many very small cells are set up, and one of the substances—usually the iron itself—begins to dissolve.

On the other hand, the corrosion of iron and steel structures that are buried underground or immersed in water can be considerably reduced by deliberately making the iron one electrode of a current-producing cell. The other electrode can be a metal such as zinc or magnesium. When the two metals are connected by a wire, hydrogen is evolved at the iron electrode, the corroding cell action that results from tiny particles of impurities in the iron is stopped, and the iron does not dissolve or corrode. The magnesium or zinc electrode is the one that dissolves. Figure 13.6 shows how such a cell can be used to protect a buried iron tank from corrosion. The "solution" is the water in the soil, which always contains some dissolved substances that allow the cell to operate. The magnesium is sacrificed to protect the more valuable iron tank. Similarly, a zinc or magnesium plate, attached to the steel hull of a seagoing ship below the water level, forms a cell in which the zinc or magnesium dissolves, protecting the steel ship from corrosion.

We have done an experiment in which we have simulated a magnesium-protected steel ship and one that is not protected. The steel of the ships' hulls is represented by rectangular strips of steel. Two new, uncorroded strips are shown in Fig. 13.7. Both strips were placed in seawater, but one was connected to a magnesium electrode (Fig. 13.8). After 35 days, the two steel strips were removed from the seawater and photographed (Fig. 13.9). As you can see, the unprotected steel is badly corroded, while the steel that formed part of the current-producing cell with the magnesium is not.

Iron or steel pipe, pails, and many iron or steel sheet-metal products are coated with a very thin layer of relatively non-corrosive zinc. Iron coated this way is called "galvanized" iron and is much less subject to corrosion than iron alone. Furthermore, if the underlying iron becomes exposed to the atmosphere as a result of scratches in the thin zinc coating, a current-producing zinc-iron cell is formed in which the zinc dissolves slowly but the iron does not. The iron is therefore protected against corrosion until enough zinc has been consumed to expose large patches of bare iron. Although zinc is more costly than iron or steel, very little is needed to give iron a thin protective coating.

Fig. 13.7 Two uncorroded steel strips before being placed in seawater.

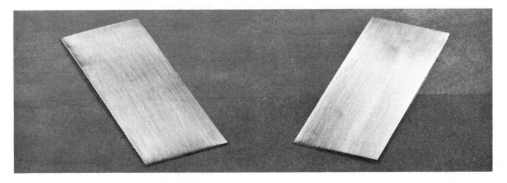

13.4 Unintentional Cells and Corrosion

Fig. 13.8 One of the steel strips is placed in the beaker of seawater on the left. The beaker on the right also contains seawater and a steel strip, but there is also a bar of magnesium, like the one lying on the board, which is connected by a wire to the steel strip. There is a current of 0.23 amp between these electrodes.

Fig. 13.9 The two steel strips after immersion in seawater for 35 days. The one on the right was protected by the magnesium electrode.

5. A man built an aluminum canoe and used copper rivets to join the parts to give the canoe a more attractive appearance. What do you think would happen if this canoe were used in salt water?

6†. Steel pails are often coated with zinc to prevent corrosion. What would happen if you used such a pail to store a solution of copper sulfate?

7. A piece of iron is part of an unintentional cell delivering a current of 10^{-3} amp. Estimate how much iron will corrode in one year.

The Motion of Electric Charge Through a Vacuum   13.5

So far we have concentrated our attention on those parts of an electric circuit where the motion of charge is related to the motion of atoms. This is the case in electrolytic cells and current-producing cells. However, it may have occurred to you that an electric current need not always be associated with the motion of atoms. You know that atoms in a solid are not free to move around, yet electric charge moves easily through a metal wire.

To get some idea of what may carry charges in a metal, we shall first describe and analyze an experiment in which charge flows through a vacuum. Figure 13.10(a) shows a vacuum tube of a kind used in radios and television sets. Nearly all the air (about 99.99999 percent) has been removed from the tube. Figure 13.10(b) shows the inside of the tube. The two most important parts are the large rectangular electrode commonly called the plate or anode, and the long, narrow cylindrical electrode called the cathode. The long zigzag wire is an electric heater which is folded

Fig. 13.10 (a) A general view of a vacuum tube. (b) Separate views of the heater, cathode, and plate.

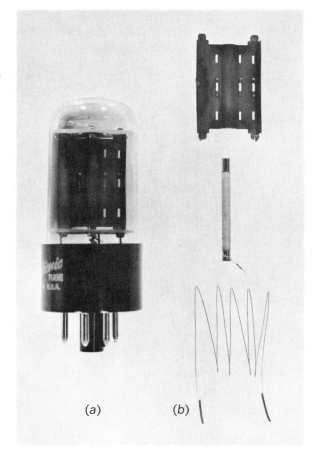

13.5 The Motion of Electric Charge Through a Vacuum

up inside the cathode and insulated from it as shown in Fig. 13.11. When supplied with current, the heater and the surrounding cathode get red hot.

We first examine whether charge can pass through the vacuum tube when the heater is not in use. Figure 13.12 shows the arrangement, including the reading of the ammeter when all connections have been made. The cathode of the tube is connected to the negative end of a 12-cell battery, and the plate is connected through an ammeter to the positive end. No current is registered by the ammeter. Next we interchange the connections of cathode and plate. As shown in Fig. 13.13, the cathode is connected through the ammeter to the positive end of the battery, and the plate is connected to the negative end. Again there is no current.

Now we repeat the connections of Fig. 13.14 and 13.15, but this time the heater inside the cathode is connected to a separate battery. There is now a current of about 0.15 amp when the hot cathode is connected to the negative terminal of the battery, but no current when it is connected to the positive terminal.

Fig. 13.11 The construction of the vacuum tube shown in Fig. 13.10 and the schematic way of representing it.

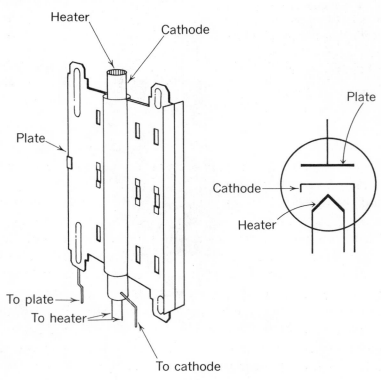

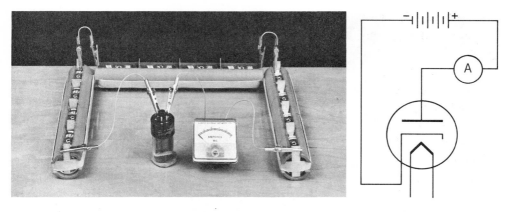

Fig. 13.12 Photograph and schematic drawing of the vacuum-tube circuit. The cathode leads to the negative terminal of the battery.

Fig. 13.13 The same circuit as in Fig. 13.12, except that the cathode is connected to the positive terminal of the battery (through the ammeter).

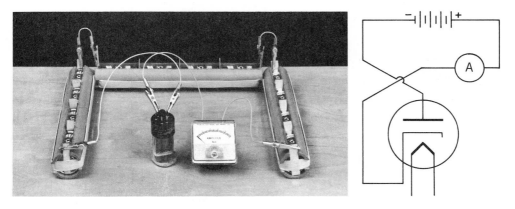

We can summarize the behavior of the vacuum tube under the conditions shown in Fig. 13.12–13.15 in the following way: When the cathode is cold, there is no current. When the cathode is heated, there is a current only if it is connected to the negative terminal of the battery.

Does this tell us anything about the direction of motion of the charge in the tube? Does the heating of the cathode enable it to receive charges coming from the plate or to emit charges which then move to the plate? To help choose between these two possibilities, think of the evaporation of water near the freezing and boiling points. Near freezing, the molecules move relatively slowly in the liquid and only a few escape per unit time. At higher temperature, near boiling, the motion becomes more vigorous and molecules leave the surface of the liquid at a high rate.

13.5 The Motion of Electric Charge Through a Vacuum

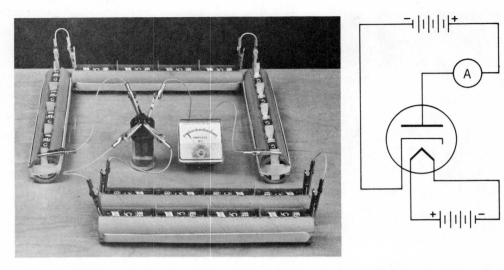

Fig. 13.14 The same as Fig. 13.12, except that the cathode is heated.

Fig. 13.15 The same as Fig. 13.13, except that the cathode is heated.

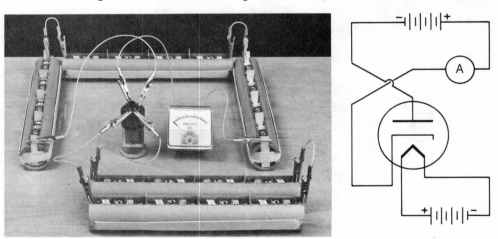

This analogy suggests that charges escape from the hot cathode and move to the anode. Does this mean that some of the atoms which make up the cathode actually carry the charge to the plate?

8† A vacuum tube has terminals which connect to the cathode, the plate, and the two ends of the heater. If you find which two go to the heater, how can you find out which terminal connects to the plate?

9 If the plate, instead of the cathode, is heated in the cases illustrated in Figs. 13.12 and 13.13, what result would you expect to observe in each case?

Electrons 13.6

We can answer the question at the end of the preceding section by massing the cathode of a vacuum tube before it is assembled and again after it has been in use for some time. But we do not even have to do that. Vacuum tubes are mass-produced so that the mass of the cathode is nearly constant for all tubes of a given type.

It will suffice, therefore, to compare the mass of the cathode of a new tube with that of a used tube. We did that and found that the mass of the cathode of the new tube and the mass of the cathode of a tube which had run at a current of 0.15 amp for 100 hours was the same: 0.224 g for both cases. What decrease in mass should we have expected if the charge flowing through the tube is carried by atoms?

There are 3.6×10^5 sec in 100 hours, so the total charge that would flow would be 0.15 amp $\times 3.6 \times 10^5$ sec $= 5.4 \times 10^4$ amp-sec. This is equal to a flow of $\dfrac{5.4 \times 10^4 \text{ amp-sec}}{1.60 \times 10^{-19} \text{ amp-sec/el. ch.}} = 3.4 \times 10^{23}$ elementary charges.

Now, to know how much mass is associated with this number of elementary charges, we must know what kind of atoms make up the surface of the cathode. Actually, the cathode surface is composed of a mixture of several oxides, since a hot surface composed of this mixture has been found to be a better emitter of charges than a pure metal surface. The lightest of the atoms in these oxides are the oxygen atoms, so, for simplicity, we shall assume that oxygen atoms carry all the charge that flows.

We shall further assume that one oxygen atom carries two elementary charges, the same charge that is required to liberate an oxygen atom in the electrolysis of water. Thus $\dfrac{3.4 \times 10^{23} \text{ el. ch.}}{2 \text{ el. ch./atom}} = 1.7 \times 10^{23}$ atoms of oxygen would leave the cathode in 100 hours. Since one atom of oxygen has a mass of 2.7×10^{-23} g, the loss in mass of the cathode should be $1.7 \times 10^{23} \times 2.7 \times 10^{-23} = 4.6$ g. This is more than twenty times the entire mass of the cathode!

The conclusion is inescapable: the charge is not carried across the vacuum tube by atoms. There must be carriers of a different kind which, from our observations, have the following properties:

(a) They come off a hot cathode connected to the negative end of a battery but not from a hot cathode connected to the positive end.

(b) They do not seem to reduce the mass of the cathode.

These carriers are called *electrons*. Since they move from the cathode

10 Suppose the cathode of a vacuum tube is made of copper. What would be the loss in mass after 100 hours at 0.15 amp if copper atoms carried the charge across the tube?

11† The mass of an electron is about $1/2{,}000$ of the mass of a hydrogen atom. How long would you have to run the vacuum tube described in Sec. 13.5 in order to show that electrons do not accumulate on the plate, increasing its mass at the expense of the cathode?

13.7 Atoms and Ions

Very early in our study of electricity we adopted the eighteenth-century idea that when an electrical device is operating, something is moving through it. We called that something the electric charge and developed ways of measuring it (Sec. 11.1). Now we are in a position to develop this idea into a more concrete model. In so doing we shall also bring the model up to date.

The first thing the model must accomplish is to tie together the motion of charge when it is accompanied by the motion of atoms and its motion when it is not. To have a definite situation in mind, consider a circuit consisting of a battery, a copper-plating cell, and a vacuum tube (Fig. 13.16). You know from your experience that in the copper cell copper dissolves at the electrode leading to the positive terminal of the battery and plates out on the electrode leading to the negative terminal. We say that charge carriers which move away from the positive terminal of a battery are *positively* charged. On the other hand, in the vacuum tube

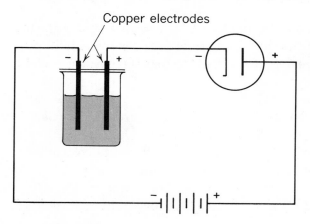

Fig. 13.16 A circuit consisting of a battery, a vacuum tube, and a copper-plating cell. The heating circuit for the cathode is not shown.

the charge carriers are electrons which move away from the cathode when it leads to the negative terminal of the battery and are negatively charged.

The problem now facing us is: how can electrons and atoms be related to account for the overall motion of charge in the circuit? This question was answered satisfactorily only in the twentieth century; and although the resulting model was based on a variety of experiments, the passage of charge through cells, metal wires, and vacuum tubes provides the main clues.

The basic features of the model are as follows: An atom is made up of a part which is positively charged and a number of negatively charged electrons. The total charge on an isolated atom is normally zero, so that the atom as a whole has no net charge; it is electrically neutral. If an atom loses an electron, it becomes positively charged. On the other hand, if an atom gains an electron, it becomes negatively charged. Charged atoms are called *ions*.

For example, when a neutral sodium atom loses an electron, it becomes a positive sodium ion. This is expressed in symbols as

$$Na \longrightarrow Na^+ + e^-$$

Similarly, when a neutral chlorine atom gains an electron, it becomes a negative chlorine ion:

$$Cl + e^- \longrightarrow Cl^-$$

Copper and zinc atoms carry two elementary charges in a solution. Thus each of them must give up two electrons to become a doubly charged ion

$$Cu \longrightarrow Cu^{++} + 2e^-$$
$$Zn \longrightarrow Zn^{++} + 2e^-$$

We shall assume that in a solution which conducts electricity the solute is in ionic form. For example, a solution of copper chloride, $CuCl_2$, conducts electricity. We can look upon the formation of the ions as the combination of the two steps

$$Cu \longrightarrow Cu^{++} + 2e^-$$
and $$2Cl + 2e^- \longrightarrow 2Cl^-$$

occurring together.* The net result is $CuCl_2 \longrightarrow Cu^{++} + 2Cl^-$.

---

*"$2Cl^-$" means two separate chlorine ions, whereas the subscript 2 in $CuCl_2$ means that in this chloride of copper there are two chlorine atoms for every copper atom.

13.8 The Motion of Charge Through an Entire Circuit

When copper sulfate, CuSO$_4$, is dissolved in water, the copper atoms separate from the sulfur and oxygen atoms as Cu^{++} ions. The negative ions, called sulfate ions, are each made up of one sulfur atom and four oxygen atoms tightly bound together. We do not know how the charges are distributed among the sulfur atom and the four oxygen atoms; therefore we shall simply write SO$_4^{--}$ for the sulfate ion to indicate that two electrons are attached to the sulfate. When an electric current passes through the copper-plating cell in Fig. 13.16, copper from the positive electrode enters the solution as Cu^{++} ions, and copper from the solution plates out on the negative electrode as neutral copper atoms. In terms of ions and electrons this suggests the following reactions:

$$\text{At the postive electrode, } Cu \longrightarrow Cu^{++} + 2e^-$$
$$\text{At the negative electrode, } Cu^{++} + 2e^- \longrightarrow Cu$$

12† A copper-plating cell and a vacuum tube were connected in series with a battery. When the heater of the tube was connected, charge moved through the circuit. Do the copper ions in the cell move away from the same battery terminal as the electrons in the tube?

13† A Daniell cell is used to run a copper-plating cell as shown in Fig. A.
    a) In which direction do the copper ions move in the copper cell? In the Daniell cell?
    b) In which direction do the zinc ions move in the Daniell cell?
    c) In which direction do electrons move in the wires?

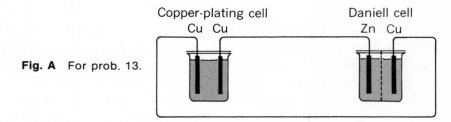

Fig. A For prob. 13.

14 When a piece of zinc is dipped into a solution of copper sulfate, zinc dissolves and copper precipitates. Describe this process in terms of atoms, electrons, and ions.

15 What is the number of elementary charges carried by an electron? Base your answer on your plating experiments and the ion-electron model of the atom.

13.8 The Motion of Charge Through an Entire Circuit

In the copper-plating cell in Fig. 13.16, positive copper ions carry charge through the solution. How is charge carried through the metal wires? From the reaction Cu $\longrightarrow$ Cu^{++} + 2e$^-$ at the positive electrode it follows that electrons are moving away from this electrode. By the same reasoning we conclude from the reaction $Cu^{++} + 2e^- \longrightarrow Cu$ at the negative electrode that electrons move from the wire to the electrode. To complete the model we must assume that the electrons can move through metals while ions cannot.

Now we can use the basic features of the model to describe what happens on the atomic level when charge moves all the way around the circuit shown in Fig. 13.16. Let us start from the negative terminal of the battery. Electrons move in the wire from this terminal to the copper electrode on the left side of the copper-plating cell. There they combine with positive copper ions which come from the solution. This produces neutral copper atoms that plate out at the negative electrode.

At the positive electrode on the right side of the plating cell in Fig. 13.16 neutral copper atoms dissolve, giving up electrons and becoming positive copper ions. At this electrode copper dissolves and replaces the copper that plates out of the solution on the left. Since copper ions are continually being removed from the solution at the left-hand electrode and at the same time being supplied to the solution at the right-hand electrode, there is a motion of positive copper ions from right to left through the solution.

The electrons given up by the neutral copper atoms that dissolve at the positive electrode of the plating cell move from the positive electrode through the wire to the hot cathode of the vacuum tube. There they "boil off" into the vacuum and move across to the plate and through the wire to the positive terminal of the battery.

This model, describing the flow of charge on the atomic level, is consistent with charge conservation in the circuit. Electrons move from the battery to a copper-plating cell. An equal number of electrons move from the copper-plating cell through the vacuum tube and then back to the battery.

What happens in the battery itself? The details depend on the kind of cells that make up the battery. But all battery cells have one property in common: they have the ability to push electrons away from one electrode and accept them at the other.

The Direction of Electric Current 13.9

The motion of charge in the copper-plating cell is particularly simple: since copper dissolves at one electrode and plates out at the other, while the rest of the cell remains unchanged, charge can only be carried across

13.9 The Direction of Electric Current

by positive copper ions. If we replace the copper cell in Fig. 13.16 by another cell consisting of two carbon electrodes and a solution of hydrochloric acid, the situation will be more complex. The electrodes themselves will remain unchanged, but hydrogen will evolve at the electrode leading to the negative terminal of the battery, and chlorine will evolve at the electrode leading to the positive terminal. This indicates that positive hydrogen ions move to the negative electrode, and negative chlorine ions move to the positive electrode. What, then, is the direction of the current in the circuit? Is it the direction of the positive ions in the solution or the direction of the negative ions in the solution and the electrons in the wire?

Apparently we could choose either direction for the direction of the current. In fact, the direction of the current is commonly taken to be the same as the direction of motion of the positive ions. The direction of the current and the direction of motion of ions and electrons is shown in Fig. 13.17.

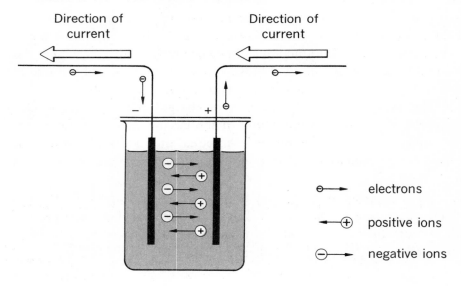

Fig. 13.17 The direction of electric current through the cell in Fig. 13.16 as related to the motion of ions and electrons.

In Fig. 13.16 the direction of the current is counterclockwise. This is the direction of the positive ions in the copper cell and is opposite to the direction of motion of the electrons in the vacuum tube and in the wires.

Notice that while outside the battery the direction of the current is from the positive to the negative terminal, inside the battery it is from the negative to the positive terminal.

For Home, Desk, and Lab

16 An experiment calls for running a Daniell cell for 15 minutes each day for a week. How would you turn off the cell at the end of each day's run so that you could continue to use the same materials for the cell?

17 A Daniell cell can be made without a divider, as shown in Fig. B.
   a) Why do the solutions not mix immediately?
   b) If the cell is allowed to stand unused, what will eventually happen?
   c) When such cells were used commercially, in the early days of telegraphy, they were never allowed to stand without producing at least a little current. Why was this?

Fig. B For prob. 17.

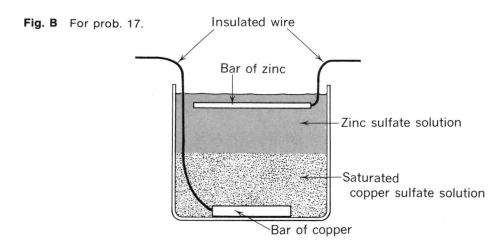

18 Will a Daniell cell produce current if both electrodes are made of zinc? Of copper?

19 Recall what happened when you dipped two zinc electrodes connected by an ammeter in a solution of copper sulfate. What do you think the ammeter would read if copper plated out uniformly and at the same rate on both electrodes without falling off?

20 You can make a simple current-producing cell by wrapping around a copper rod a paper towel soaked in a solution of ammonium chloride and sprinkled with manganese dioxide powder. A piece of heavy-duty aluminum foil can serve as both a case and an electrode for your cell. Will this cell light a flashlight bulb?

For Home, Desk, and Lab

21 If you were to make an iron-copper cell, which electrode would you expect to be negative?

22 A circuit like that in Fig. 13.14 is wired up with an additional ammeter placed in the heater circuit. Before the battery at the top is connected, the heater is turned on and a current of 0.60 amp is read on the meter in the heater circuit. When the battery at the top is connected, the meter at the top in the cathode-plate circuit reads 0.16 amp. What reading would you now expect to find on the heater-circuit meter?

23 You have worked with hydrogen cells, plating cells, ammeters, light bulbs, and vacuum tubes. If you place these things in a circuit, which ones will enable you to tell the "+" terminal of an unmarked battery? How would you tell?

24 In some types of vacuum tube, the heater is used as the cathode. Such a tube is shown in Fig. C. The ammeter is of a type in which the zero position of the needle is in the center of the scale, and the needle can move in either direction. S_1 and S_2 are switches.

What do you predict you would observe if the circuit were operated under each of the following conditions?

	Switch S_1	Switch S_2	Observation
A	Open	Closed upward	
B	Closed	Closed upward	
C	Closed	Closed downward	

25 Suppose you have a cell consisting of two copper plates and a solution of sulfuric acid. What do you predict will happen if you pass a current through this cell for some time?

26 A student using the evidence so far in the course proposed that ions form only when water is present. Does the evidence presented in Sec. 6.9 support his argument?

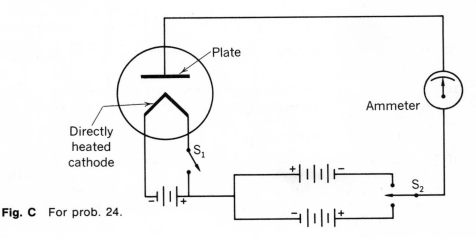

Fig. C For prob. 24.

Heat 14

You know from daily experience that electric irons, toasters, stoves, and light bulbs all get hot when electric charge flows through them. Apparently a flow of charge can produce heat in certain parts of a circuit. Before we investigate the relation between heat and the flow of electric charge, we need to find a way to measure heat.

Throughout your work in *Introductory Physical Science* you used an alcohol burner to heat substances or a water bath to cool them, and in many experiments you measured the temperature of the substance during the process. You were not really concerned with the number of burners you used or the time it took to heat a sample to the desired temperature. Similarly, you did not worry about the exact amount of water you used in a water bath. Yet these things can be very important. Consider the two beakers containing water at room temperature shown in Fig. 14.1. When

Fig. 14.1 Both beakers initially contain water at room temperature and are heated by identical burners. In which beaker will the water boil first?

14.1 Experiment: Heating Different Masses of Water

you use identical alcohol burners, it does not surprise you that it will take longer to get the beaker with more water to boil than the one containing less. You also know from daily experience that if you were in a hurry to boil the water, you could use two burners under one beaker. We can change the temperature of two different samples of matter by the same amount, but we may have to heat one longer than the other. Or, <u>if we wish to produce the same temperature change in the same length of time, we may have to heat one sample more</u> strongly than the other.

There is, therefore, another aspect to heating an object which is not described by the change in temperature of the object but rather by the number of alcohol burners and the time they were in use. Clearly, two identical burners heating two test tubes filled with water supply twice as much of something as does one burner heating one tubeful for the same length of time; but this something certainly is not the temperature, since the temperature change is the same in each case. This something is the quantity of heat that the burners transfer to the water. <u>It is not surprising that it takes twice the quantity of heat to produce the same temperature change in twice the mass of water</u>. Raising the temperature of twice the mass means, in terms of the atomic model, making twice as many atoms move faster.

Does it take the same amount of heat to raise the temperature of 1 g of water 1°C from 20°C to 21°C as it does from 40°C to 41°C? Does it take the same quantity of heat to raise the temperature of 1 g of water 1°C as it does to raise the temperature of 1 g of iron 1°C? To answer these questions, we must first find a way to compare quantities of heat.

Experiment
14.1 Heating Different Masses of Water

Figure 14.2 shows apparatus that can be used to compare quantities of heat. Suppose the two small electric heaters are identical. Then, whenever one heater supplies heat to a given mass of water to produce a given temperature change, the other heater will supply the same amount of heat to whatever is in its cup.

To find out if the two heaters you have are identical, you can use them to heat 50 g of water in each of the two insulating cups. When these cups are used to measure quantities of heat, we call them *calorimeters*. Measure the water temperature in both cups as accurately as you can before you start heating.

Experiment: Heating Different Masses of Water   14.1

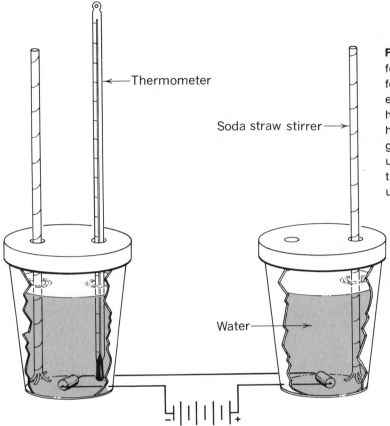

Fig. 14.2 Two insulating styrofoam cups with insulating styrofoam covers, and a pair of small electric heaters used to supply heat to water in the cups. The heaters are turned on and off together. The two soda straws are used as stirrers to thoroughly mix the water in the cups before measuring the temperature.

To measure the temperature changes accurately, you will need a sensitive thermometer. Examine the thermometer supplied for this experiment. What fraction of a degree can you read on the scale with reasonable confidence?

Now choose one of the cups as a "standard" and connect the heaters to the battery as shown in Fig. 14.2. Stir the water in the standard calorimeter continually until its temperature rises 2.00°C (as nearly as you can tell). Then you can disconnect the heaters from the battery and, after stirring, read the temperature of the water in the other cup as accurately as possible. What is the temperature rise? Did the two heaters produce the same quantity of heat?

Use the two heaters to investigate the temperature rise of different quantities of water when a given quantity of heat is supplied. Starting each time with 50 g of water at room temperature in your standard calorimeter, make several runs, using a different mass of water at room temperature in the other calorimeter. To be sure that the same quantity of heat is delivered in different runs, disconnect the battery when the temperature of the water in the standard calorimeter rises as close to 2.00°C as you can read.

How does the temperature change of a mass of water greater than 50 g compare with that of the standard? How about a mass of water less than 50 g?

How do the products of mass and temperature change compare for the different masses of water that you tried?

14.2 The Calorie

We can show that differences in the initial temperature of a sample of water do not affect its temperature rise when the water is supplied with a fixed quantity of heat. We have done this, using apparatus that reduces heat losses to the surroundings to a much lower level than the apparatus you used in the last experiment. Our new apparatus is larger and can hold more water. We heat water in it using an electric immersion heater like the kind used to boil water in a cup for instant coffee. Two such electric immersion heaters and tanks are shown in Fig. 14.3. During our experiments with this apparatus, both tanks held 2,000 g of water, and both were stirred continuously as they were being heated simultaneously. The temperatures in both tanks were measured with thermometers that could be read to 0.02°C.

To begin the experiment, the water in each tank was at room temperature. Several runs were made in which there were 10°C temperature rises in the tank used as a standard. In each of these runs there was also a 10°C rise in the second tank, thus confirming that each heater supplied the same amount of heat in the same time. Then several runs were made in which the starting temperature was always room temperature for the standard tank but different for the other tank. In each case heating was stopped when the temperature rise for the standard tank was 10°C.

For all runs the temperature rise in both tanks was the same within 3 percent. Table 14.1 shows the experimental results. Note that not only were the temperature changes for the two tanks the same, but the products of mass of water and temperature change are also the same within 3 percent.

We have used the apparatus just described to do the experiment you did in the last section. The results were just as precise as those shown

Fig. 14.3 A scaled-up version of the apparatus in Fig. 14.2 in which the heat losses are very small. The heaters are on the table in front of the styrofoam tanks. When immersed in the tanks they are turned on and off together by the switch shown between the tanks. The pegboards support electric stirrers and the thermometers. The battery between the tanks operates the stirrers. The insulating covers are not shown.

Table 14.1 Results of Supplying the Same Amounts of Heat to the Same Mass of Water at Different Initial Temperatures

Initial Temperatures (°C)		Temperature Changes (°C)		Mass × Temperature Change (g × °C)	
Tank 1	Tank 2	Tank 1	Tank 2	Tank 1	Tank 2
Room Temp.	5.0	10.0	10.2°C	2.00×10^4	2.04×10^4
Room Temp.	15.0	10.0	10.2°C	2.00×10^4	2.04×10^4
Room Temp.	25.0	10.0	10.1°C	2.00×10^4	2.02×10^4
Room Temp.	35.0	10.0	9.9°C	2.00×10^4	1.98×10^4

in Table 14.1. For different masses of water between 1 kg and 3 kg, supplied with a fixed quantity of heat, we found that, like the experiment summarized in Table 14.1, the product of mass and temperature change was constant within 3 percent.

We conclude, therefore, that whether we vary the mass of water heated or the initial temperature, the product of mass and temperature change is constant when a fixed amount of heat is supplied to water. Therefore we can use the product of mass and temperature rise as a measure of the heat transferred to water, regardless of the initial temperature. In fact, we shall use as a convenient unit of heat the quantity of heat needed to raise the temperature of 1 g of water 1°C. This unit is called the *calorie* (cal). As you can see, in the experiments we have just described, the water in the tanks was supplied in each case with $(2{,}000 \times 10.0)$ cal $= 2.00 \times 10^4$ cal. Whenever we raise the temperature of water, we can find out how much heat (in calories) was supplied to the water by calculating the product of the mass of the water in grams and the temperature rise in degrees centigrade:

$$\text{Heat added (cal)} = \text{mass of water (g)} \times \text{temperature change (°C)}.$$

1† A thousand grams of water is heated with an immersion heater. The temperature of the water rises from 10°C to 25°C. How many calories have gone into the water?

2 In the experiment described in this section, what would have been the temperature rise if 4,000 g of water had been heated in the second tank while 2,000 g of water in the first tank was heated 10°C?

3 How many calories are needed to heat 1 g of water from its freezing point to its boiling point?

Experiment
14.3 Heating Different Substances

We now come back to the second question raised at the beginning of the chapter: Do equal masses of different substances rise in temperature by the same amounts when supplied with the same quantity of heat? You can answer this question by using the heaters of Expt. 14.1 to heat equal masses of water and another liquid.

Using 50 g of water in your "standard" calorimeter and 50 g of cooking oil (its density is 0.9 g/cm³) in the other calorimeter, find the temperature rise of the oil when the water changes temperature by 2.00°C. How many calories of heat were absorbed by the cooking oil? How much heat was required to raise the temperature of 50 g of cooking oil 1°C? How much heat is required to raise the temperature of 1 g of cooking oil 1°C?

Heat Capacity; Specific Heat   14.4

In the experiment you have just done, you found a value for the quantity of heat absorbed by a 50-g sample of cooking oil when its temperature rises by 1°C. The quantity of heat absorbed by any sample of matter when its temperature rises 1°C is called its *heat capacity*. Clearly, the heat capacity of a sample of matter depends on its mass: the greater the mass of a sample of matter, the greater the amount of heat it must absorb to increase in temperature by 1°C. For example, it takes 10 cal of heat to raise a 10-g sample of water 1°C, and the heat capacity of this sample is therefore 10 cal/°C. But a 20-g sample of water requires 20 cal to increase its temperature 1°C, so its heat capacity is 20 cal/°C, twice that of a 10-g sample.

In Expt. 14.3 you compared the heat capacity of a sample of cooking oil with that of an equal mass of water. They were not the same. Apparently the heat capacities of samples of different substances having the same mass is a characteristic property that we can use to distinguish one substance from another. For convenience, we choose to compare the heat capacities of samples of different substances whose masses are all 1 g, just as in the case of density where we compare the masses of samples of matter whose volumes are all 1 cm^3.

The heat capacity of 1 g of a substance is called its *specific heat*. The specific heat of a substance is, therefore, the quantity of heat that must be absorbed by 1 g of the substance to raise its temperature 1°C. The specific heat of the cooking oil you used in Expt. 14.3 is the heat capacity in cal/°C of the sample divided by its mass in grams. Therefore, specific heat is measured in (cal/°C)/g, which is the same as (cal/g)/°C.

The specific heat of water is 1.0 (cal/g)/°C, since we have defined the calorie as the quantity of heat needed to raise the temperature of 1 g of water by 1°C. From Table 14.1 we can conclude that the specific heat of water is practically independent of temperature, since equal quantities of heat supplied to a fixed mass of water produced very nearly the same temperature change, regardless of the initial temperature.

To determine heat capacities and specific heats more accurately than you were able to in Expt. 14.3, we used the two electric immersion heaters and styrofoam tanks shown in Fig. 14.3. The experiment was performed with 2,000 g of water in one tank and 2,000 g of glycol (permanent antifreeze) in the other. One of the heaters was placed in the water and allowed to run until the temperature of the water had increased by 10.0°C. This took 305 sec. The same heater was then placed in the glycol and allowed to run until the temperature of the glycol had increased by 10.0°C.

14.4 Heat Capacity; Specific Heat

This took 190 sec. Since the same heater was used with both samples, heat was added to the glycol at the same rate as it was added to the water, but the time required to change the temperature of the glycol by 10.0°C was (190 sec)/(305 sec) = 0.62 of the time required to change the temperature of the same mass of water by 10°C. Since the amount of heat supplied by the heater was proportional to the time of heating, the 2,000 g of glycol absorbed only 0.62 as much heat in warming 10°C as the same mass of water did in warming 10°C. To raise the temperature of 2,000 g of water 10°C required 2.00×10^4 cal; but only $0.62 \times 2.00 \times 10^4 = 1.24 \times 10^4$ cal was needed to raise the temperature of the same mass of glycol 10°C. The quantity of heat absorbed by the sample of glycol in warming 1°C (its heat capacity) was

$$\frac{1.24 \times 10^4 \text{ cal}}{10°C} = 1.24 \times 10^3 \text{ cal/°C}.$$

Since the sample of glycol has a mass of 2.00×10^3 g, the specific heat of glycol, the heat required to warm 1 g by 1°C, is

$$\frac{1.24 \times 10^3 \text{ cal/°C}}{2.00 \times 10^3 \text{ g}} = 0.62 \text{ (cal/g)/°C}.$$

Table 14.2 gives the specific heat of some common substances. Notice that the specific heat of water is much higher than that of all the other common substances except hydrogen. This is generally true; very few substances have a specific heat as large as that of water.

Table 14.2

Substance	Specific Heat (cal/g)/°C	Substance	Specific Heat (cal/g)/°C
Aluminum	0.215	Oxygen	0.22
Copper	0.092	Ice	0.50
Iron	0.108	Water	1.00
Lead	0.031	Water vapor	0.48
Silver	0.0563	Moth nuggets (liquid)	0.30
Hydrogen	3.4	Calcium chloride	0.16
Nitrogen	0.25	Olive oil	0.47

Very precise experiments indicate that the specific heat of all substances is affected slightly by the temperature. In our experiments, however, we shall be working in so narrow a range of temperatures that the small change of specific heat with temperature is not important.

4† A sample of a liquid absorbs 200 cal of heat in warming from 20°C to 25°C. What is its heat capacity?

5† What is the heat capacity of 2,000 g of water?

6 A sample of a metal absorbs 2.8 cal of heat in warming from 20.0°C to 25.6°C.
   a) What is its heat capacity?
   b) What additional information must you have to determine the specific heat of the sample?

7† An electric immersion heater like that described in Sec. 14.2 supplied 1.33×10^3 cal of heat to a sample of mercury. The temperature of the mercury rose from 20.0°C to 40.0°C.
   a) What was the heat capacity of the sample of mercury?
   b) The mass of mercury was 2.00×10^3 g. What is the specific heat of the mercury?

8 Equal masses of olive oil and water are heated in identical containers for the same length of time on the same hot plate. If the temperature of the water increases 5°C, what will be the approximate temperature change of the olive oil?

Experiment
Heat Lost by a Substance in Cooling 14.5

Fig. 14.4 Warm and cold substances can be mixed in a styrofoam cup with little heat lost to or gained from the surroundings. The cover is made from the bottom of a styrofoam cup. Such an arrangement is called a calorimeter.

We have defined the specific heat of a substance as the quantity of heat needed to raise the temperature of 1 g by 1°C. But what if we cool the substance? Will 1 g of water, for example, lose as much heat when its temperature is lowered by 1°C as it gains when it rises by 1°C? You can answer this question by mixing cool water with warm water, determining the heat lost by the warm water in cooling by calculating the heat gained by the cool water in warming. You will have to insulate the warm water from its surroundings as much as possible so that it will not cool off by warming the surrounding air. You can do this quite effectively by placing the warm water in a calorimeter consisting of a stryofoam cup placed in a beaker (Fig. 14.4) and pouring room-temperature water into it.

From the mass and rise in temperature of the cold water, calculate how much heat it gained. Assume that no heat leaks to the surroundings, so that all the heat from the water that cools goes into warming up the cold water. Using this assumption, how much heat was lost by the water that cools?

How much heat was lost by 1 g of the water that cooled? How much heat was lost by 1 g of this water in cooling 1°C? What do you conclude about the heat lost per gram per °C when water cools?

9† A hundred grams of water at 60°C is mixed with 200 g of water at 0°C. The temperature of the mixture is 20°C. How much heat did the cold water gain? How much heat did the hot water lose?

Experiment
14.6 Heat Capacity and Specific Heat of a Solid

If you have a solid and want to measure its heat capacity, you can apply the knowledge you gained in the last experiment.

Try to find the heat capacity of a metal cylinder by heating it and placing it in cool water. The cylinder you will use is a specially made one that you will use in heat experiments in later chapters. It contains two holes. Fill the large one with plasticine to keep out water, and consider this plasticine as part of the cylinder. (Its specific heat is nearly the same as the cylinder's.) To heat the cylinder, hang it by a thread in a beaker of boiling water. What is the temperature of the boiling water? How can you be reasonably sure when the temperature of the cylinder is the same as that of the boiling water?

When you are reasonably sure the cylinder is at the temperature of the boiling water, remove it, and quickly put it into the container of cool water.

How much heat was gained by the cool water when you put the hot metal into it? How much heat was lost by the metal? Through what temperature range did it cool? What is the heat capacity of the cylinder? What is the specific heat of the metal? What metal do you think the cylinder is made of?

10 Some metal washers, whose total mass is 200 g, are heated in boiling water and then placed in 100 g of water at 20°C. The final temperature of this mixture is 25°C.
   a) What is the specific heat of the metal?
   b) What metal might the washers be made of?

For Home, Desk, and Lab

11 Suppose you heat 100 g of room-temperature water in a small beaker and 1,000 g of room-temperature water in a large beaker until the temperature in each is 60°C.
  a) Do the water molecules in the large beaker now move as fast as do those in the small one?
  b) Did the beakers of water gain the same amount of heat?

12 The graphs in Fig. A are cooling curves for 50-g samples of water in four different containers. The room temperature was 22.0°C.
  a) Which container is the best to use as a calorimeter?
  b) How much heat was lost from container 4 during the period from 0 to 10 minutes? From 70 to 80 minutes?

13 A hundred grams of water at 90°C and 100 g of water at 50°C are placed in separate but identical containers. Which container has the greater rate of cooling? (Use Fig. A.)

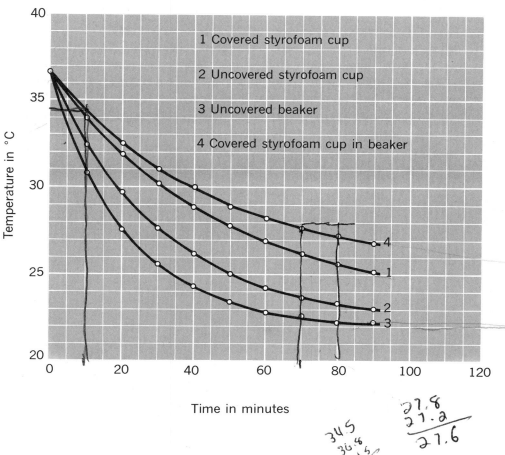

Fig. A For probs. 12 and 13.

For Home, Desk, and Lab

14 A certain heater coil is known to supply 1,000 cal/min. If this coil is placed in 500 g of water in an insulated container,
  a) How many calories will the coil supply in 2 min?
  b) What will be the temperature rise in 2 min?

15 Apparatus similar to that shown in Fig. 14.3 is used to heat 1,000 g of water and 1,000 g of another liquid. When the heaters are turned off, the temperature of the water has risen 10°C. The temperature of the second liquid has increased 30°C.
  a) What is the heat capacity of the second liquid?
  b) What is its specific heat?

16 The specific heat of zinc is 0.0925 (cal/g)/°C. What is the heat capacity of 4.00 g of zinc?

17 a) What mass of aluminum has the same heat capacity as 20 g of water?
  b) What mass of iron has this heat capacity?

18 The specific heat of oil is about half that of water. Why did defenders of castles in the Middle Ages pour hot oil rather than hot water over the walls onto attacking soldiers?

19 Which liquid, water or glycol, would be better to use in a hot-water bottle?

20 In Expt. 3.14 you plotted a graph of temperature against time for water being heated to the boiling point.
  a) From this graph would you predict that it takes less time to raise by 5°C the temperature of water at 90°C or water at room temperature?
  b) Does your answer agree with the fact that the specific heat is very nearly constant between 0°C and 100°C?

21 Hundred-gram samples of aluminum, copper, and lead are placed in boiling water. Each sample is then removed and placed in a container with 100 g of water at 20°C. Which metal will cause the greatest temperature change in the cool water? (See Table 14.2)

Heat and Electric Charge 15

Now that we have an understanding of heat and can measure heat capacities, we can investigate the relationship between the quantity of heat generated in an electric heater and the quantity of charge that flowed through it.

Experiment

The Heating Effect of a Flow of Charge   15.1

How is the quantity of heat generated in a heater related to the quantity of charge and the current that passes through it?

You can determine the charge that flows through the heater by using an ammeter and clock. You can find the quantity of heat produced in a heater by placing it in a calorimeter and measuring the rise in temperature.

The calorimeter consists of the aluminum cylinder whose heat capacity you measured in Expt. 14.6, and two styrofoam blocks (Fig. 15.1). Although styrofoam is a good heat-insulating material, some heat leaks out. This loss can be reduced by starting below room temperature and stopping the run when the temperature has risen an equal amount above room temperature. In this way the heat gained from the surroundings while the cylinder is below room temperature will partly compensate for the heat lost to the surroundings while the cylinder is warming up above room temperature. You can use a plastic bag containing an ice cube to

15.1 Experiment: The Heating Effect of a Flow of Charge

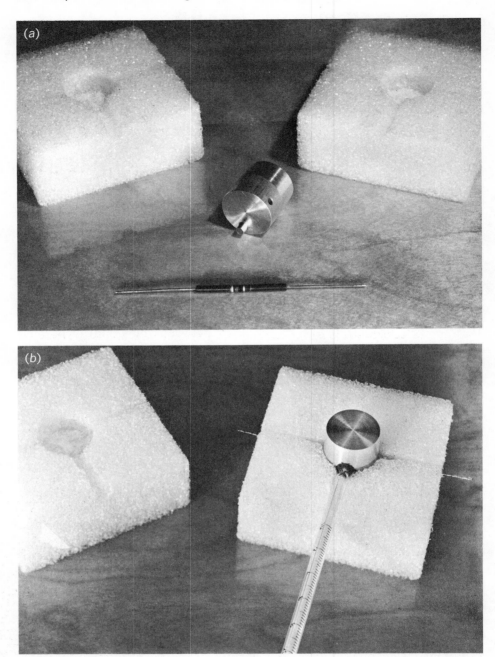

Fig. 15.1 The calorimeter and heater used in Expt. 15.1 are shown in (*a*). The small heater in the foreground fits into the small hole in the aluminum cylinder. A thermometer can be inserted in the large hole, its bulb kept in close contact with the cylinder by means of some soft plasticine. The heater and thermometer are seen in place in (*b*).

cool the cylinder (Fig. 15.2). Although starting a run below room temperature reduces the heat loss, it does not eliminate it, so it is best not to exceed a temperature rise of about 10°C.

After adjusting the initial temperature of the cylinder with a heater in place, connect the circuit, shown in Fig. 15.2, to a battery of eight flashlight cells and record the time that charge flows through the heater. Note that the temperature continues to rise a bit after the circuit is disconnected; record the maximum temperature.

How much heat was produced by the small heater? How much charge flowed through it?

Repeat the experiment using other heaters, collecting data for different temperature changes of at least 4°C but not more than 10°C. You can use your data to draw a graph of the heat produced in calories as a function of the charge in amp-sec that flowed through the heater. Should the point representing zero calories and zero charge lie on the line you draw on the graph? What do you conclude from the graph?

Fig. 15.2 After the aluminum cylinder has been cooled below room temperature with the ice cube in the plastic bag shown on the left, the styrofoam cover is placed on the calorimeter and the apparatus is connected as shown. Before putting the cover in place, be sure to wipe off any water that condenses on the cylinder with a paper towel. If you get the temperature too low, you can warm the cylinder the desired amount by touching it briefly with your hand. **CAUTION:** The small heater should not be connected to the battery unless it is imbedded in the cylinder. If charge flows through it while it is exposed to the air, it may burn out.

15.2 Experiment: Heat Produced as a Function of Number of Flashlight Cells

What was the heat produced in calories when one amp-sec of charge flowed through the heater in each run? We shall refer to this quantity as the heat per unit charge. Does the heat produced per unit charge depend on the current?

1† a) What is the heat capacity in cal/°C of an aluminum cylinder whose specific heat is 0.215 (cal/g) / °C and whose mass is 20.0 g?
  b) How much heat flows into the cylinder if its temperature rises 4.0°C?

2 Why was it suggested in the experiment above that the temperature rise be kept between 4°C and 10°C?

3† A student noticed the following variations in the temperature readings on his thermometer in Expt. 15.1 after disconnecting the battery. He made a reading in °C roughly every 20 seconds and the results were: 30.00, 30.95, 30.95, 30.90, 30.85, 30.75, 30.70. Which of these values best represents the temperature to be used in calculating the heat released by the heater?

4† In Expt. 15.1, what is the effect on the final temperature reading of starting 6°C above room temperature?

Experiment
15.2 Heat Produced as a Function of Number of Flashlight Cells

In Expt. 15.1 you investigated the relation between heat and charge flow for a fixed number of cells. In this experiment you will use the same apparatus to find out how the heat produced by a unit charge flowing through a heater depends on the number of flashlight cells used to move charge through the heater.

You can make the same measurements in the same way as in Expt. 15.1. This time, however, make several runs with varying numbers of flashlight cells starting with eight and ending with four cells. As you use fewer cells, you will need to use longer times to obtain a reasonable change in temperature.

How much charge flowed through the heater in each run? Draw a graph of the heat produced per unit charge as a function of the number of cells used, and compare its shape with the graphs of your classmates. What conclusion can you draw from the graph? How much heat is produced per amp-sec by one cell?

5† Suppose you used an alcohol burner and raised the temperature of 50 g of water by 10°C in 100 sec.

a) If you allowed 0.25 amp to flow through a heater to produce the same number of calories in 100 sec, about how many cells would you require?
b) How many cells would be needed if the current was 0.1 amp?

The Voltmeter 15.3

Experiment

The preceding experiment leaves no doubt that the quantity of heat you can get from a heater depends not only on the charge but also on the number of flashlight cells in the battery to which it is connected. Suppose now that someone covered up a battery of cells so that you could not see how many cells made up the battery. There is a way, besides counting the number of cells in a battery, of determining the amount of heat per unit charge a battery will supply. An instrument called a voltmeter can give you directly the heat per unit charge available from a battery.

A voltmeter has a terminal marked "+". This terminal is connected to the positive terminal of the battery. With a voltmeter and your battery, make measurements from which you can draw a graph of voltmeter readings (voltage) in units called volts as a function of the number of cells.

What is the relationship between the readings on the voltmeter and the number of cells? Does any charge flow through the voltmeter? What can you do to find out?

6† You can buy batteries which give a voltmeter reading of 6.0 volts, and others which give a reading of 90 volts. If you know that these batteries are made up of a number of flashlight cells, how do you explain the difference in their voltmeter readings?

Heat Produced as a Function of Voltage and Charge 15.4

You are now in a position to draw some far-reaching conclusions by combining the results of the three experiments you did in this chapter. In Expt. 15.2 you found that the heat produced per unit charge in a heater is proportional to the number of cells in the battery. In the last experiment you found that also the voltmeter reading or voltage across the heater

15.4 Heat Produced as a Function of Voltage and Charge

is proportional to the number of cells in the battery. To put it differently, if we double the number of cells, we double both the voltage across the heater and the heat per unit charge produced in the heater. Leaving the number of cells out of the discussion, we see now that doubling the voltage across the heater doubles the heat produced per unit charge. Tripling the voltage triples the heat per unit charge. In general, then, the heat per unit charge is proportional to the voltage.

In Expt. 15.1 you saw that for a fixed number of cells the heat produced is proportional to the quantity of charge flowing through the heater. We now know from Expt. 15.3, The Voltmeter, that this result will also hold for a fixed voltage.

How will the heat produced in the heater change if, for example, we double the voltage across it and triple the quantity of charge passing through it? Doubling the voltage will double the heat, and tripling the charge will triple the heat. The total quantity of heat will, therefore, increase $2 \times 3 = 6$ times. We can state this conclusion in a general way as follows: The heat produced in an electric heater is proportional to the product of the charge and the voltage or

$$\text{Heat} = (\text{constant}) \times \text{charge} \times \text{voltage}.$$

The numerical value of the constant depends on the units in which we measure heat, charge, and voltage. We did an experiment resembling Expt. 15.1 to determine this value; however, we also measured the voltage supplied to the heater by the battery. Instead of the aluminum cylinder we used water in a styrofoam cup for the calorimeter. The heater was submerged in the water, and we measured the temperature rise of the water. Here are the results of one run:

Mass of water in the calorimeter	100 g
Change of temperature	7.0°C
Heat generated in the calorimeter	$7.0 \times 100 = 700$ cal
Current	2.15 amp
Time of run	120 sec
Electric charge passing through the heater	$2.15 \times 120 = 258$ amp-sec
Voltage across the heater	11.3 volts

Since heat = (constant) × charge × voltage,

$$\text{Constant} = \frac{\text{heat}}{\text{charge} \times \text{voltage}} = \frac{700 \text{ cal}}{258 \text{ amp-sec} \times 11.3 \text{ volts}}$$

$$= 0.24 \text{ cal/amp-sec-volt}$$

We have now a powerful tool to find the heat generated in any of the heaters we used by measuring only the electrical quantities, charge and voltage. From these quantities, the heat produced is:

$$\text{Heat (cal)} = 0.24 \left(\frac{\text{cal}}{\text{amp-sec-volt}}\right) \times \text{charge (amp-sec)} \times \text{voltage (volts)}$$

7 An electric immersion heater can operate from a 12-volt automobile battery. It draws a current of 10 amp. About how long will it take to bring to a boil enough water to make a cup of tea?

8 Use the graphs you drew in Expt. 15.2 and 15.3 to make a graph of heat per unit charge produced as a function of voltage.

Electrical Work   15.5

The product (charge) × (voltage) appears so often that it is useful to give it a name. We shall refer to it as the *electrical work*, or simply as *work*. Since we measure charge in amp-sec and voltage in volts, it is natural to define the unit of electrical work as the work done by 1 amp-sec passing between two points in a circuit with 1 volt across them. This unit is called a *joule*:

$$1 \text{ joule} = 1 \text{ amp-sec} \times 1 \text{ volt}$$

The relation which summarizes the experiments with the heater can now be written as

$$\text{Heat (measured in cal)} = 0.24 \text{ cal/(joule)} \times \text{work (measured in joules)}$$

The calorie was defined in Sec. 14.2 as the quantity of heat needed to raise the temperature of 1 g of water 1°C. We can equally well define a unit of heat as that obtained from one joule of electrical work. In this case, the number of units of heat in joules that we get from an electric heater is the same as the number of joules of electrical work done on the heater:

$$\text{Heat (in joules)} = \text{Work (in joules)}$$

Suppose a current of 5 amp passes through a heater for 10 seconds when there is a voltage of 2 volts across it. The electrical work done and also the heat produced in the heater is then 5 amp × 10 seconds × 2 volts = 100 joules. In calories, the heat produced is 0.24 cal/joule × 100 joules = 24 calories.

15.5 Electrical Work

The calorie is a larger unit than the joule, since it takes fewer of them to express the same amount of heat:

$$24 \text{ cal} = 100 \text{ joules}$$
$$\text{and } 1 \text{ cal} = \frac{100}{24} = 4.2 \text{ joules.}$$

Notice that although both heat and electrical work can be expressed in joules, they are measured in different ways. In Expt. 15.2 you found the quantity of heat produced in an aluminum cylinder from its heat capacity and change in temperature. The electrical work you found from readings on a clock, an ammeter, and a voltmeter.

Multiplying the voltage by the current will give the amount of electrical work done in one second when an electric heater is running. This product, which equals the electrical work done per second, is called the electric power.

$$\text{Power} = \frac{\text{work done}}{\text{time}} = \frac{\text{voltage} \times \text{current} \times \text{time}}{\text{time}} = \text{voltage} \times \text{current}$$

The unit of electric power expressing the rate at which electrical work is done is the *watt** which is equal to 1 joule/sec. Most common electrical appliances have the power rating stamped on them—for example, a 100-watt or a 60-watt light bulb. The total electrical work done when an appliance is operated is the product of the power and the time. This is sometimes expressed in watt seconds (joules) or more commonly in kilowatt hours (kw-hr). A kilowatt hour is $1000 \times 3600 = 3.6 \times 10^6$ watt-seconds (joules).

9 A battery supplies 10 volts to a heater. The current through the heater is 0.3 amp, and charge flows through the heater for 60 sec.
  a) How much electrical work is done?
  b) How many joules of heat are produced in the heater? How many calories is this?

10 One electric heater delivered 400 joules of heat to a cup of water, as measured by a voltmeter, ammeter, and clock. Another heater delivered 120 cal to a cup of water, as determined by the mass of water heated and the temperature rise. Which heater delivered the most heat?

11† A small heater draws a current of 0.5 amp for 500 sec and produces 1,000 cal of heat. What is the heat per unit charge in (a) cal/amp-sec (b) joules/amp-sec? (c) What would be the reading of a voltmeter connected across the heater?

---

*James Watt was responsible, in the early part of the last century, for increasing the power obtainable from a steam engine of a given size. In the course of his work with steam engines he introduced the first unit of power, the horsepower.

12† In the vacuum tube in Fig. 13.10 the heater draws 0.6 amp when 12.6 volts are applied. How many joules of heat are produced by the heater each second?

13† The voltage applied to the terminals of a 60-watt light bulb was measured to be 115 volts. What was the current through the bulb?

14† How many joules of electrical work are done on a 100-watt light bulb that burns for 3 hours?

Electrical Resistance 15.6

You have connected different heaters to a battery and measured the current through them with an ammeter. You found that for a given battery (a fixed voltage), the current was different through different heaters. In fact, the heaters you used are purposely made so that the current through them is different when the same voltage is applied to them. The electrical property of the heaters that relates the current through them to the voltage applied to them is called electrical resistance. We shall define it as the ratio of the voltage to the current. That is:

$$\text{Resistance} = \frac{\text{voltage}}{\text{current}}$$

The electrical resistance of an object depends on the substance of which it is made. Compare the current through an aluminum wire, a tin wire, and a thin piece of wood, all having the same dimensions and all having the same voltage applied, and you will find the largest current in the aluminum wire, while hardly any current will be detected in the wood. For objects of the same dimensions the electrical resistance of wood is much larger than that of tin, which in turn is larger than that of aluminum.

Yet the resistance of an object does not depend only on the material of which it is made, or on its dimensions. The ratio (voltage)/(current) depends on the temperature of the object and in many cases also on the voltage itself. However, for many conductors, particularly metals at constant temperature, the ratio (voltage)/(current) is independent of the voltage. Figure 15.3 shows the current through a piece of iron wire as a function of the voltage across it. You can see from the graph that when the voltage across the wire is doubled, the current also is doubled; when the voltage is tripled, so is the current. Or, in short, the current

15.6 Electrical Resistance

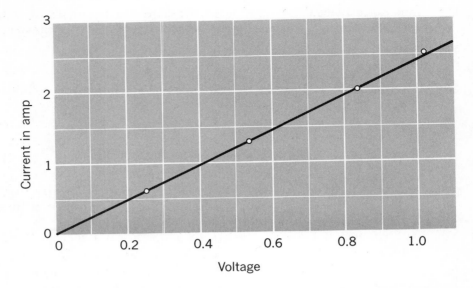

Fig. 15.3 The current through a piece of iron wire as a function of the voltage across it. The maximum voltage applied was not enough to increase the temperature of the wire appreciably.

is proportional to the voltage. The proportionality between voltage and current is known as Ohm's law, after the German physicist Ohm (1787–1854).

For an object that obeys Ohm's law, we can measure the voltage and current once, calculate the ratio (which is the resistance), and mark it on the object. When connected to a different voltage across it at the same temperature, we can calculate the current that will flow through it from the relation: current = (voltage)/(resistance). For example, a piece of copper wire with 10 volts across it is found to carry a current of 2 amp. The resistance, therefore, is 10 volts/2 amp = 5 volts/amp. The resistance of copper is constant for a given temperature; hence, if we connect the same piece of wire to 50 volts, there will be a current of 10 amp through it, since the constant ratio is 5 volts/amp.

The unit of resistance 1 volt/amp is called an *ohm*. The heaters you used so far obey Ohm's law if the temperature stays constant. The resistance is indicated by the color band marked on the heater. For example, the yellow, violet, black heater has a resistance of 47 ohms. The major use of such heaters in radios, television sets, or other electronic devices is to maintain required voltages across various parts of a circuit. Because they limit or resist the flow of charge, they are usually referred to as "resistors." We called them "heaters" simply because we used them for heating.

Both metal wires and the resistors you have seen have a constant resistance only as long as their temperature does not change. The currents recorded in Fig. 15.3 are so small that the iron wire remained practically at the same temperature. Figure 15.4 shows how the current changes as a function of voltage when the currents get larger and the iron wire heats up. The proportionality between current and voltage no longer holds; doubling the voltage from 5.0 to 10.0 volts raises the current only from 5.8 to 6.7 amp. Ohm's law does not hold when the temperature changes. The resistance of iron increases with temperature. In fact, this is true of most conductors. However, there are some, like carbon, whose resistance decreases when their temperature rises.

Fig. 15.4 Current as a function of voltage for the same wire that was used to make the graph in Fig. 15.3, but with the maximum voltage ten times as great—so large that the wire became almost red hot. Over the range shown, the current is not proportional to the voltage, and Ohm's law does not hold beyond about one volt. The segment of the graph from zero to one volt is the same as the graph shown in Fig. 15.3.

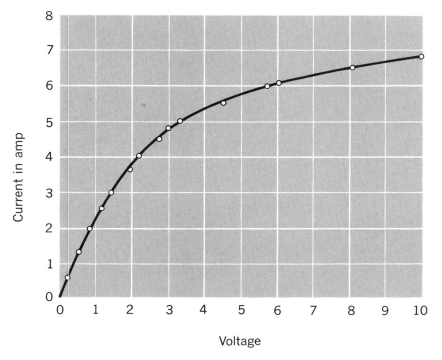

15.7 Experiment: Heat Produced in a Daniell Cell

15† What is the work when 3.0 amp flows for 20 seconds through a 5.0 ohm resistor?

16† A resistor immersed in water is used as a heater. When connected for 10 seconds, it supplied 10 joules of heat, while the current was 0.5 amp.
  a) What was the voltage across the resistor?
  b) What is the heater's resistance?

Experiment
15.7 Heat Produced in a Daniell Cell

In the experiments you have done in this chapter, you have assumed that all the heat produced was generated in the heaters. This was very nearly true in these experiments, but there are situations in which it is far from true. Under certain circumstances a significant quantity of heat is generated in the battery that is used to move charge through a circuit. In this experiment we shall look for heat in a cell that produces current. We shall use a Daniell cell instead of a flashlight cell, because it is easier to measure temperature changes in it.

You can use the same Daniell cell that you used before, with about 50 cm³ each of copper sulfate solution and of zinc sulfate solution at room temperature in the separate compartments of the cell. An ammeter, a voltmeter, and the Daniell cell are connected as shown in Fig. 15.5.

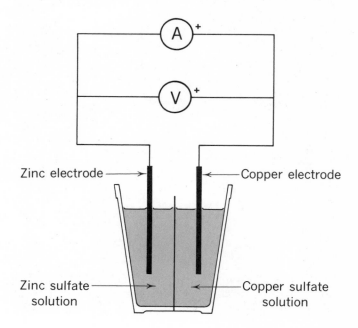

Fig. 15.5 The circuit shows a Daniell cell connected to an ammeter and a voltmeter. The ammeter has a low resistance and it "short-circuits" the cell as effectively as a short length of copper wire.

Before completing the circuit with the ammeter, record the starting temperatures of the solutions in the Daniell cell. What is the voltage of the Daniell cell just before completing the circuit?

You can now complete the circuit by connecting the ammeter, and let it operate for about 10 minutes. What are the voltage and current of the cell while it is operating? After opening the circuit, stir the solutions and read the temperature of each.

How did the voltage supplied by the cell to the ammeter compare with the cell voltage when it was not delivering current? Did the flow of charge produce any heat in the cell?

———   ———   ———

The Daniell cell which you have just used in this experiment produces such a small temperature change that a quantitative experiment is not possible with the tools at your disposal. We therefore constructed a different cell to measure quantitatively the heat produced by a unit charge. We made the cell narrower, so as to reduce the volume of solution without reducing the surface area of the electrodes, and used a better heat-insulating material.

The cell was operated twice, once connected to a heater in a calorimeter (Fig. 15.6), and once short-circuited by clamping the plates together (Fig. 15.7). The results are summarized in Table 15.1.

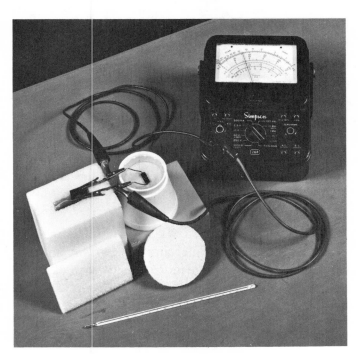

Fig. 15.6 A Daniell cell well insulated against heat losses connected to a heater in a calorimeter. The calorimeter cover and the thermometer have been removed to show the small heater. The cell contains very small volumes of the two solutions, so that the temperature rise obtained will be large enough to read precisely. The voltage supplied by the cell is measured with a more accurate voltmeter than the one you have used.

15.7 Experiment: Heat Produced in a Daniell Cell

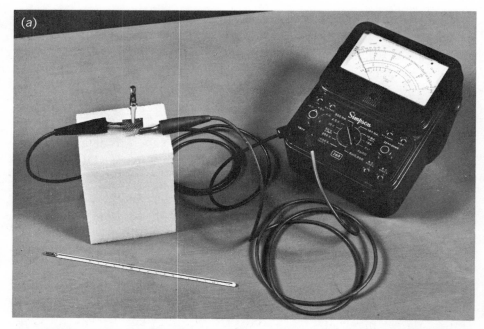

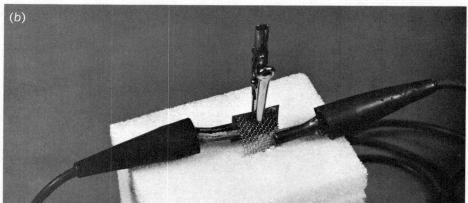

Fig. 15.7 (a) The same Daniell cell as shown in Fig. 15.6 but with the heater and calorimeter removed and electrodes of the cell clamped together to form a short circuit. Note that the voltage between the terminals is zero. (b) How the electrodes are clamped together.

In a Daniell cell we do not need to measure the current and the time to find the charge; we can find the charge more directly by measuring the change in mass of one of the electrodes.

As you can see, with the Daniell cell connected to a heater, some heat (330 joules) was produced outside the cell, in the calorimeter. At the same time 240 joules was produced inside the Daniell cell itself.

Table 15.1

	Cell Connected to Heater	Cell Short-Circuited
Change in Mass of Zn Electrode	0.17 g	0.44 g
Total Charge Flowing	5.0×10^2 amp-sec	1.3×10^3 amp-sec
Voltage Across Terminals of Cell	0.68 volt	0 volt
Heat Produced in Cell	240 joules	1,400 joules
Heat Produced in Calorimeter	330 joules	0 joules
Total Heat Produced	570 joules	1,400 joules
Total Heat/Unit Charge	1.1 joules/amp-sec	1.1 joules/amp-sec

When the heater was disconnected and the cell was short-circuited by clamping the electrodes together, all the heat (1,400 joules) was produced inside the Daniell cell.

Table 15.1 shows that the total charge flowing, the voltage at the terminals of the cell, and the total heat produced were all quite different in the two cases, but the heat produced by a unit charge was the same. In other words, a given mass of zinc dissolving in a Daniell cell produces a specific quantity of heat regardless of whether all the heat is generated inside the cell or part of it is produced outside the cell. In particular, the quantity of heat produced when one atom of zinc goes into solution in the Daniell cell is independent of the details of the circuit. This conclusion turns out to be valid also for other cells: the quantity of heat generated by one atom going into solution depends on the cell and not on what is connected to the cell. However, whether the heat appears mostly outside the cell or inside it depends on the current, and hence on the details of the circuit.

The highest voltage across a cell is observed when the cell is providing no current (except a very small current to the voltmeter). This is called the "open-circuit voltage." For a Daniell cell this voltage is 1.1 volts. Note that this is the same as the heat per unit charge in joules/amp-sec in Table 15.1. This observation shows again the relation:

$$1 \text{ volt} = 1 \frac{\text{joule}}{\text{amp-sec}}$$

17 The open-circuit voltage of a battery of 10 Daniell cells is 11 volts. When it is connected to a resistor its voltage drops to 8 volts. A current of 0.5 amp is drawn for 20 sec.
 a) How much heat is generated in the resistor?
 b) How much heat is generated in the battery?

18 In an experiment like the one described in Sec. 15.7, one-half the total heat produced was found in the heater calorimeter. What could be done, using the same Daniell cell, to get more heat into this calorimeter than into the Daniell cell?

For Home, Desk, and Lab

19 If all of the heaters supplied for Expt. 15.1 had lost their markings after the experiment, how could you determine which was which?

20 A student argued that if the "+" terminal of the ammeter in Expt. 15.1 were connected to the "−" terminal of the battery, the aluminum cylinder should cool, because the pointer of the ammeter would read "less than zero." Is he right? (He pointed out that cooling is produced in a refrigerator by electric current!)

21 Three calorimeters like the one you used in Expt. 15.1 are connected in series to a battery. The heaters in the calorimeters are identical. The metals are different in the various calorimeters, but the mass of metal in each is the same. In each case, what would you predict about
a) the current in each heater?
b) the heat generated in each calorimeter?
c) the heat produced by a unit charge in each?
d) the temperature rise in each?

22 Figure A is a graph which resulted from Expt. 15.3
a) How can you explain the results?
b) Check your explanation, using your battery.

23 A heater like the one you used in Expt. 15.2 is connected in series with four cells.
a) If you let this heater run for 15 seconds, how much heat would you expect to be produced?
b) If you now used eight cells in series and arranged the time to give the same charge, how much heat would be produced?
c) Now you let the circuit run five times as long as in (b). How much heat would you expect to be produced?
d) Do your answers agree with the statement that the heat produced is proportional to the product of charge times voltage?

24 If you double the voltage across a heater of constant resistance, will the heat produced in a given time also double?

25 Twenty grams of water in a plastic-cup calorimeter are heated by an electric heater through which 0.30 amp of current flows for 100 seconds from a five-cell battery.
a) What would be the rise in temperature?
b) What would you expect the temperature rise to be if 10 cells were used?

For Home, Desk, and Lab 81

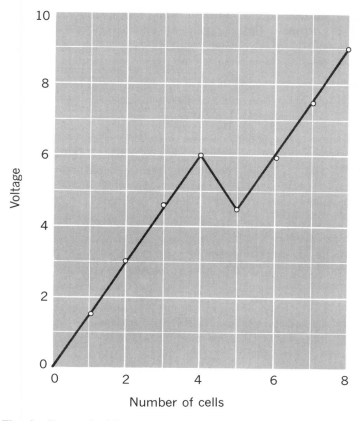

Fig. A For prob. 22.

Fig. B For prob. 28.

26. One horsepower equals 746 watts. Light bulbs, therefore, could be rated in terms of horsepower instead of watts. What would be the horsepower rating of a 60-watt bulb? At how many horsepower would you rate a 2,000-watt electric stove burner?

27. If electrical work is charged for at the rate of 5 cents/kw-hr, (a) how much does it cost to operate a 100-watt bulb for one evening (3 hours)? (b) What is the cost per joule of electrical work?

28. Fig. B shows a long nichrome wire, of the kind used in many electric heaters, supported on a pegboard. When the wire is connected to the battery, what do you predict about (a) the heat generated in equal segments (AB, BC, etc.) of the wire? (b) the voltage across each segment? (c) the sum of the voltages across all segments? (d) How does the resistance of a wire depend on its length?

29. Find the resistance of the heaters you used in Expt. 15.1 by connecting an ammeter, a voltmeter, and each of the heaters in turn to an eight-cell battery as shown in Fig. C. *CAUTION:* be sure the heaters are in the aluminum cylinder when current is supplied to them! Do you need to use the styrofoam insulation?

30. Using the circuit shown in Fig. C, measure the current flowing through the lowest-resistance heater you used and the voltage applied to the heater when different numbers of cells are used in the battery. Draw a graph of current as a function of voltage.

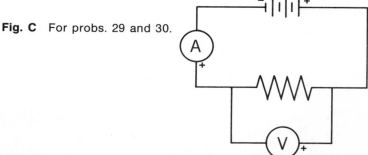

Fig. C For probs. 29 and 30.

31. What can you determine about a copper-plating cell if you know only (a) current in the cell and time the charge flowed? (b) current and voltage applied to the cell? (c) voltage and time? (d) current, voltage, and time?

32. What would you do to see if a copper-plating cell follows Ohm's law?

33. Voltmeters have widely differing resistance from ammeters. What would you do to find out which have the larger resistance?

34. How many Daniell cells would you have to connect together to operate a 9-volt transistor radio?
    a) How would you connect them?
    b) Which electrode in the Daniell cell is the "+" terminal?

For Home, Desk, and Lab

35 A Daniell cell supplies current to a heater. What is the largest quantity of heat that can be generated in the heater if 0.17 g of zinc is dissolved in the cell? Would you choose a high- or a low-resistance heater? (See Table 15.1.)

36 A Daniell cell was balanced against some gram masses on an equal-arm balance. Do you expect the cell to remain balanced if it is short-circuited until all the zinc is dissolved?

37 A voltmeter alone is connected to a battery of Daniell cells and reads 12.1 volts. Now a heater is added. Which of the following quantities can you determine without further information? For which do you need additional information, and what information is it?
a)   The heat per unit charge generated in the heater alone.
b)   The sum of the heats per unit charge generated in the heater and in the battery.
c)   The heat generated in the heater in 10 sec.
d)   The total heat generated per second.

38 A mass of 0.109 g of zinc dissolved in a Daniell cell. Using the data in Tables 12.1 and 12.2, determine how much charge passed through the circuit.

16 Where Is the Heat?

16.1 Heat Produced by an Electric Motor

In the preceding chapter we saw that the heat produced in a resistor by a unit charge is proportional to the voltage across the resistor. In particular, when the heat is measured in joules and the electric charge in ampere-seconds, the heat per unit charge is numerically equal to the voltage. We can show that this result holds independently of the current for a variety of heaters, whether or not they are resistors which obey Ohm's law, whether or not their temperatures remain constant.

Does the relation

$$\text{Heat (in joules)} = \text{electrical work (in joules)}$$
$$= \text{charge (in amp-sec)} \times \text{voltage (in volts)}$$

hold under all conditions? Let us try to answer this question by measuring the quantity of heat that is produced in an electric motor under different operating conditions. Figures 16.1, 16.2, and 16.3 show three different

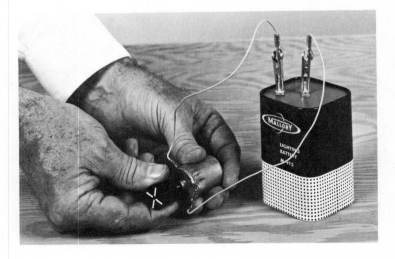

Fig. 16.1 A small electric motor with a cylindrical black hard-rubber drum fastened to its shaft. The motor is connected to a battery, but it does not run because the drum is firmly held.

Heat Produced by an Electric Motor **16.1**   85

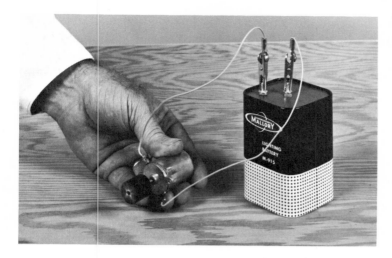

Fig. 16.2 The motor shown in Fig. 16.1, running freely.

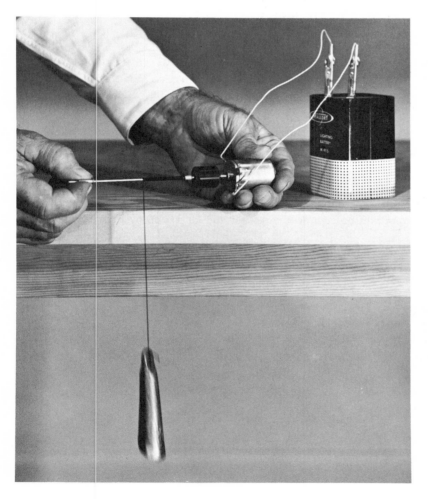

Fig. 16.3 The motor lifting an object.

16.2 Experiment: The Heat Capacity of an Electric Motor

conditions under which we can connect a motor to a battery and let electric charge pass through it. In Fig. 16.1 the shaft of the motor is held fixed so that it cannot turn even when a current passes through the motor. In Fig. 16.2 the motor is allowed to run freely, and in Fig. 16.3 it is lifting an object.

Let us consider the three cases separately, beginning with a motor whose shaft is held fixed.

1† How many joules of heat can be produced in a heater in 3 minutes by a current of 0.50 amperes if the voltage across it is 1.2 volts?

Experiment
16.2 The Heat Capacity of an Electric Motor

Figure 16.4 shows the inside, rotating part of a motor like the one you will be using. When the motor is connected to a battery, the electric charge passes through the coils of insulated copper wire. When the motor shaft is held fixed, we would expect the motor to act like a heater made of copper wire. In fact, this is the case. The only effect of the passage of charge through it will be the generation of an amount of heat in joules equal to the electrical work, the product of charge times voltage.

Fig. 16.4 The rotating part of the motor shown in Fig. 16.1. Coils of insulated copper wire can be seen in the slots in the large iron cylinder.

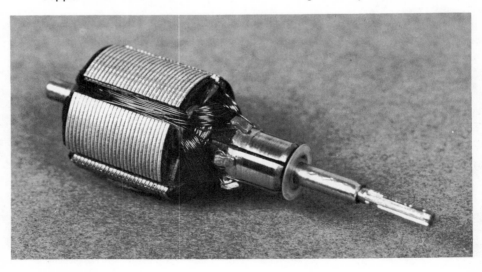

Experiment: The Heat Capacity of an Electric Motor **16.2**

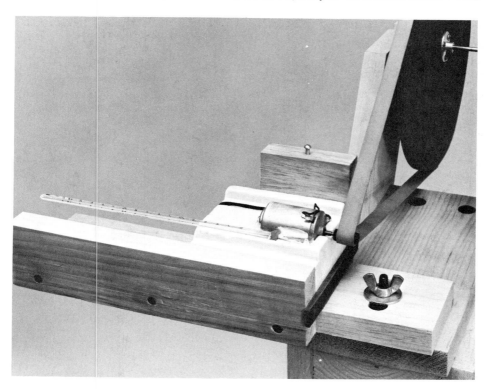

Fig. 16.5 The top of the insulating box has been removed to show the position of the motor and thermometer. You can keep the bulb of the thermometer in close contact with the motor by placing a little plasticine in the small metal tube that holds the thermometer. Be sure that the thermometer is positioned so that its scale is facing toward you.

If you measure the temperature rise of a non-running motor when supplied with a known amount of electrical work, you can calculate its heat capacity. You can then use this value to measure the heat produced in the motor when it is free-running and when it is lifting an object.

You can measure the rise in temperature of the motor with a thermometer kept in close contact with the side of the motor by plasticine. To reduce heat losses to the surroundings, the motor is enclosed in a box well insulated with styrofoam. Figure 16.5 shows the insulated box (with its top off) holding the motor and with a thermometer in place.

You can use the circuit shown in Fig. 16.6(*a*) to pass charge through the fixed motor. In this circuit eight flashlight cells are used in the battery; four 2-cell batteries are connected in parallel as shown in Fig. 16.6(*b*). These parallel-connected batteries will provide essentially the same current as a single battery made up of only two series-connected cells, but

16.2 Experiment: The Heat Capacity of an Electric Motor

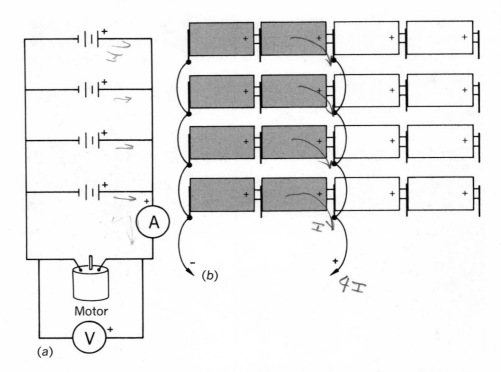

Fig. 16.6 (a) The circuit used in measuring electrical work and, thus, the heat produced in the motor. (b) The arrangement and connections of cells for the battery shown in (a).

they will take four times as long to "run down" because each two-cell parallel branch supplies only one-fourth of the total current. Therefore, the voltage and current will stay more nearly constant during a run, and this will result in a more accurate determination of the electrical work.

Attach a clothespin to the shaft so that it cannot rotate (Fig. 16.7). Be sure the temperature of the motor is within a few tenths of a degree of room temperature. Part of the styrofoam cover can be removed to allow the motor to come to room temperature.

When you are sure that the temperature of the motor is constant, put the styrofoam cover in place. Record the temperature (to the nearest 0.05°C) and connect the motor to the battery for about 1 minute, recording the average voltage and current and the time.

After you disconnect the battery, record the temperature of the motor every 30 seconds until you are sure it is falling. (It will take the motor a few minutes to reach its maximum temperature.) What is the maximum temperature reached by the motor after it is disconnected? Using the data obtained by all the members of the class, make a histogram and decide on the best value for the heat capacity of the motors. (The motors are all very nearly identical.)

Fig. 16.7 A clothespin attached to the shaft of the motor prevents it from turning.

2   When voltage is applied to a motor and it is not allowed to run, it acts like a heater. From your voltage and current readings, calculate the resistance of the motor you used.

3   The same kind of cells are used in each of the circuits shown in Fig. A. Each cell provides 1.5 volts. What is the voltage across the resistor and the current flowing through each cell in each of the circuits?

Fig. A For prob. 3.

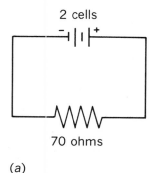

(a)

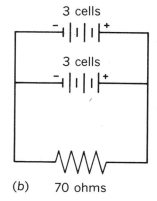

(b)

16.3 Experiment: A Free-running Motor

4. In Expt. 16.2, suppose the thermometer was not kept in close contact with the motor with plasticine. How would this affect the results of the experiment?

Experiment
16.3 A Free-running Motor

When the shaft of a motor is free to rotate, is the heat produced in it equal to the electrical work done?

The voltage and current requirements of a motor when its shaft is rotating are quite different from the voltage and current needed when the shaft is not allowed to rotate. For this reason the battery in this experiment consists of four 4-cell batteries connected in parallel (Fig. 16.8). Adjust the rubber belt connecting the large pulley and the motor-shaft pulley so that it is well aligned and will not slip off the pulleys when the motor is running. To reduce friction, you can place a drop of oil on the two large pulley-shaft bearings and on the motor-shaft bearing. It is wise to run the motor for a few seconds to see that the belt stays on the pulleys. When the motor returns to room temperature, you can proceed as before and run it for about one minute, recording the exact time, the average current and voltage, and the initial and maximum temperatures.

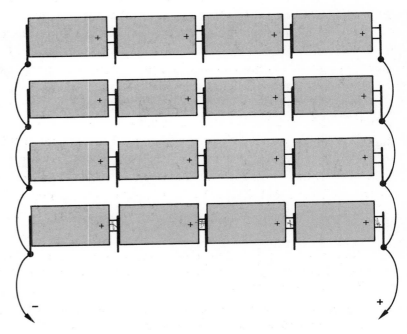

Fig. 16.8 The battery used in Expt. 16.3, 16.4, and 16.5.

Using the best value for the heat capacity of the motor from the preceding experiment, compare the heat generated in the free-running motor with the electrical work. Does the electrical work equal the heat generated in a free-running motor?

5   If a motor heats up while running, how can it run steadily without burning out?

6† The same amount of electrical work is done on two identical well-insulated motors. One motor shaft turns freely while the other is held fixed. How will the temperature changes in the two motors compare?

Experiment
A Motor Lifting an Object   16.4

The next condition under which we want to measure the heat produced in a motor is when it is lifting an object. The same battery is used in this experiment as in the previous one.

The object to be lifted consists of a bucket with 3 kg of mass added to it and a wooden dowel (used to make the bucket more rigid). Slip a small loop of nylon string over the small peg near the center of the large pulley. (Do not fasten the loop tightly to the peg; it must be able to slip off easily in the next experiment.) Now clip a hook to the handle of the bucket and attach the other end of the nylon string to the hook so that the string is just long enough to allow one turn of the large pulley to barely lift the bucket off the floor.

When the motor is at room temperature, you can connect it to the battery. When the bucket has been lifted as far as it will conveniently go, immediately turn off the motor and hold the bucket at this level. (Be sure to record the average current and voltage during the run, and the time of the run.) Now you can unhook and remove the bucket and then record the maximum temperature reached by the motor.

How does the heat generated compare with the electrical work?

7† Two identical motors are operated at the same voltage, and the same charge passes through both. One motor turns freely while the other lifts an object. Which motor will get hotter?

8   Can you think of a place in the mass-lifting apparatus where heat is generated by friction and not taken into account?

Experiment
16.5 Where Is the "Missing Heat"?

In the first two experiments with the electric motor the arrangement of the apparatus is the same at the end of the experiment as it was at the beginning. At the end of the last experiment, however, a bucket had been raised to a higher position above the floor and a significantly smaller amount of heat was generated by a given amount of electrical work. Is the lifting of the bucket related to the "missing heat"? What happens if the bucket is allowed to fall back slowly to the floor? To find out, you can repeat the last experiment, but when the bucket reaches its top position do not touch it, but disconnect the circuit and let the bucket return to the floor.

How does the total heat generated when the bucket was both raised and allowed to fall in the same operation compare with the electrical work?

For Home, Desk, and Lab

9. Suppose you performed Expt. 16.2 with the same time, current, and voltage, but using another motor of larger mass. How would this affect
    a) the change in temperature of the motor?
    b) the total heat absorbed by the motor?

10. Plot a graph of the current through a free-running motor against the voltage applied using from one to four flashlight cells. Does a free-running motor obey Ohm's law? Do the same for a stalled motor. Does a stalled motor obey Ohm's law?

11. Suppose you used a battery of Daniell cells in one of your experiments with the motor, and the current in each cell was so small that almost no heat was generated in the battery. If the electrical work was 100 joules, how much zinc was dissolved in the battery? (Use Table 15.1.)

12. The shafts of two motors are connected together. A flashlight bulb is connected to the terminals of one motor. When a battery is connected to the terminals of the other motor, both motors run and the lamp lights. How would you expect the electrical work from the battery to compare with the heat produced in the motor to which it is connected?

13. If you used a motor to lift a mass 1 meter and the temperature of the motor changed T°C, what would you expect the change in temperature to be (a) if you lifted the same mass twice? (b) if you lifted the same mass once through 2 meters?

14 How do you think the missing heat would compare with the missing heat in Expt. 16.4, if (*a*) you had lifted less mass? (*b*) you had lifted the same mass through a greater height?

17 Potential Energy

17.1 Gravitational Potential Energy and Thermal Energy

In Chapter 15 you found that the quantity of heat generated in a heater depends only on the electrical work done, that is, on the product of charge and voltage. Moreover, when both electrical work and heat are measured in joules, they are numerically equal.

In Chapter 16 you compared, under different conditions, the quantity of heat generated in an electric motor by a fixed amount of work. We can now sum up the results of the motor experiments in the following way:

A free-running motor acts like a heater; the heat generated equals the electrical work.

$$\text{Heat (joule)} = \text{work (joule)}$$

In a motor lifting an object, the same amount of electrical work yields smaller quantities of heat:

$$\text{Heat (joule)} < \text{work (joule)}$$

However, the missing heat is substantially recovered by allowing the object to fall slowly through the same distance it was lifted, turning the motor as it falls. The missing heat is not lost after all, since by lifting the object, we have created a situation which allows us to get it back.

Thus, in the motor experiments electrical work can be used either to lift an object without changing the object's temperature, or to raise the temperature of a motor without changing the motor's position. Furthermore, we can trade the rise in position of an object for a rise in temperature of the motor. For this reason the same word is used to describe the two changes: in both cases we say that the *energy* of the object

Experiment: Gravitational Potential Energy as a Function of Mass **17.2**

increases. When a body is lifted, its *gravitational potential energy* increases. When the temperature of a body rises, its *thermal energy* increases. The increase in thermal energy is equal to the quantity of heat added to the body. It makes no difference whether the heat flowed in from the outside or was generated in the body by an electric current. We shall measure the changes of both kinds of energy in the same units, mostly in joules.

The idea that lifting an object and raising its temperature are in a sense equivalent is a generalization based on the results you obtained in the motor experiments. There were many chances for experimental errors in the measurements you made, and no doubt, your class results were quite scattered. In most cases not all the missing heat was recovered.

We can reduce errors by designing the apparatus more carefully and using better instruments, but we can never eliminate them entirely. Therefore, no matter how accurate our measurements are, we are making a guess when we extrapolate our results to ideal conditions in which there is no heat loss to the surroundings and no error in measuring temperature, current, etc. Therefore, it takes a great deal of intuition to draw valid and far-reaching conclusions from actual measurements, particularly if they are not very accurate. Experiments similar to the ones you made were first performed by the English physicist James Joule (1818–1889). His early results were no more accurate than yours, yet Joule realized their implications; his conclusions contain the idea that electrical work can produce heat in a motor directly, or indirectly by lifting an object and letting it fall.

1† How does the increase in thermal energy of 100 g of water when it is heated from 20°C to 50°C compare with the decrease in thermal energy when it is allowed to cool back down to 20°C?

2† A piece of hot iron is dropped into a calorimeter containing cold water. How does the change in thermal energy of the iron compare with the change in thermal energy of the water?

Experiment
Gravitational Potential Energy as a Function of Mass   17.2

We could use the motor and objects of different mass to investigate the relationship between the gravitational potential energy and the mass of the falling object. All we have to do is let different masses fall a given distance and graph the rise in temperature of the motor as a function of the mass of the falling object. However, there are two disadvantages in using the

17.2 Experiment: Gravitational Potential Energy as a Function of Mass

motor. First the motor conducts heat so poorly that we must wait several minutes for its temperature on the outside to reach its maximum value. Second, its temperature rise is small even for falling masses as large as 3 kg, which makes it difficult to determine the temperature rise accurately. To eliminate these disadvantages, we need a device which conducts heat better than the motor and which has a smaller heat capacity so that its temperature rise will be larger.

A small, solid aluminum cylinder will do the job. It provides us with a way to convert gravitational potential energy to thermal energy directly without using the motor. To do this we make use of friction. You know that when your hands are cold, you can warm them by rubbing them together vigorously. In this experiment a string pulled by a falling object will generate heat by rubbing on an aluminum cylinder.

Figures 17.1 and 17.2 show an aluminum cylinder clamped in a block of wood so that it cannot rotate. The nylon string wrapped around the cylinder rubs against it as the string is pulled by a falling bucket. The thermometer inserted into the cylinder measures the rise in temperature resulting from the heat produced by the rubbing string.

Fig. 17.1 Apparatus for measuring the heat produced by a slowly falling bucket. The string holding the bucket is looped around a fixed aluminum cylinder and then passes down over a steel guide rod on the left and is hooked to a counterweight shown near the bottom of the photograph. The bucket was part of the way down in its descent when the photograph was taken.

Experiment: Gravitational Potential Energy as a Function of Mass 17.2

Fig. 17.2 A close-up of the top of the apparatus shown in Fig. 17.1. The thermometer projects from the rear of the cylinder.

The end of the string that is not connected to the bucket is attached to a small counterweight, whose purpose is to hold the string tight against the aluminum cylinder and control the speed at which the bucket falls.

We are interested, in this experiment, in finding the relation between the mass of the falling bucket and the heat produced in the cylinder. Therefore, the mass that falls is the mass in the bucket plus the mass of the bucket itself minus the mass of the counterweight. We have to subtract the mass of the counterweight because it is lifted during a run, and we want the net mass that falls.

Before making a measured run it is necessary to adjust both the number of turns of string around the cylinder and the mass of the counterweight so that the falling mass descends slowly. It should take between 5 and 15 seconds to fall to the bottom. With one bag of sand in the bucket and two or three turns of string around the cylinder as shown in Fig. 17.3, try the different counterweights singly or together until the bucket falls at a satisfactory rate.

Lift the bucket so that its bottom is 1.00 m from the floor. Then let it fall.

17.2 Experiment: Gravitational Potential Energy as a Function of Mass

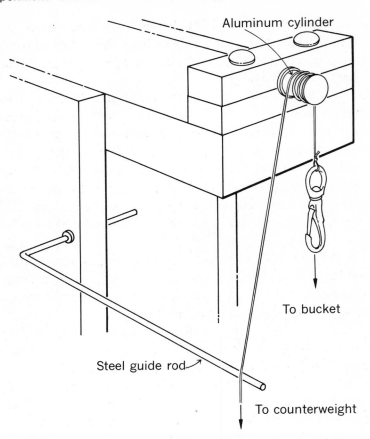

Fig. 17.3 As the bucket slowly falls, the loops of string move along the cylinder. If they move to the back end of the cylinder before the bucket reaches the floor, some of the loops may become tangled. To prevent this, wind the string around the cylinder clockwise as shown in the drawing and then slide the loops along the cylinder so that they form a single layer and are up against the rim on the front of the cylinder.

The aluminum cylinder loses heat rapidly to the surroundings, and it is important to keep this loss to a minimum. This can be done by starting a run with the cylinder below room temperature, as you did in the experiments in Chapter 15 in which you used an aluminum cylinder as a calorimeter. In those experiments you could choose an initial temperature, and stop the heating when the temperature had risen as far above room temperature as it started below room temperature. However, in this experiment you do not know what the temperature rise will be. To determine how much to cool the cylinder, you first make a run starting from room temperature. Suppose you observe a temperature rise of about 3.0°C. If you then cool the cylinder to 1.5°C below room temperature and make another run, the heat lost to and gained from the surroundings will very nearly cancel each other, although the new temperature rise will be slightly greater than 3.0°C. Why?

With two kilograms in the bucket, make your first run starting with the aluminum cylinder at room temperature. From the measured temperature rise of this run, decide on the appropriate initial temperature. After cooling the cylinder to this temperature, make a second, corrected run.

Repeat the experiment several times, using different masses and adjusting the number of turns of string around the cylinder and the combination of counterweights to give a slow descent of the bucket. Should you start at the same temperature when making runs with different masses?

From the temperature changes and the heat capacity of the aluminum cylinder, 20.0 joules/°C, calculate the increase in thermal energy of the aluminum. Now make a graph of the increase in thermal energy as a function of the net mass that falls.

We reasoned, as a result of the motor experiments in Chapter 16, that a decrease in gravitational potential energy can be converted into an increase in thermal energy. Therefore, in this experiment we can use the increase in thermal energy of the cylinder as a measure of the decrease in gravitational potential energy of the falling bucket. What do you conclude about the relationship between decrease in gravitational potential energy of the bucket and its mass when it falls through a given height? What is the decrease in gravitational potential energy when an object of 1.00 kg mass falls 1.00 m?

3   If you increase the mass of the counterweight, what effect will it have on the speed of descent of the bucket you used in this experiment?

4   Suppose you arranged the apparatus in such a way that it took the bucket 10 minutes to reach the floor. What rise in temperature would you expect to find?

5   Why can the mass of the string and the hooks attached to it be neglected when you determine the net mass that descends?

6   How would an increase in the heat loss from the aluminum cylinder to the air affect the graph you drew in this experiment?

7   How would the results have been affected if
    a) you began with the temperature of the aluminum cylinder a degree or two above room temperature?
    b) you forgot to subtract the mass of the counterweight from the mass of the bucket and its contents?

Experiment
17.3 Gravitational Potential Energy as a Function of Height

To investigate the dependence of the change in gravitational potential energy of a body on the height through which it falls, you can use the same apparatus and procedure you used in the preceding experiment. This time you change the distance the bucket falls instead of the mass. For the mass, use the largest one you used in Expt. 17.2, so that for small distances of fall, you will get a measurable temperature rise.

Make a graph of the change in thermal energy of the aluminum cylinder as a function of the distance the bucket falls. What do you conclude about the dependence of the change in gravitational potential energy on the change in the vertical position of an object?

In the preceding experiment you found the loss in gravitational potential energy when 1.00 kg dropped 1.00 m. Using this number and the general conclusions you drew from these two experiments, write an equation for the change in gravitational potential energy for any mass changing its vertical position by any distance.

8† From the results of Expt. 17.2, we know that the change in gravitational potential energy of an object is proportional to its mass; in Expt. 17.3 we see that it also is proportional to the distance the object falls. How do you expect the gravitational potential energy to change if you double both the mass and the distance it falls?

9† Which gains more gravitational potential energy—an 8-kg object lifted 1.0 m, or a 6-kg object lifted 2.0 m?

17.4 The Change in Gravitational Potential Energy over Different Paths

When you investigated the change in gravitational potential energy as a function of mass and height, the object moved down vertically. Will the change in gravitational potential energy be the same if we keep the change in height the same but let the object move down an incline? When the object moves down various inclines which form increasing angles with the vertical, the distance it travels also increases (Fig. 17.4). Does this affect the quantity of heat that is generated and therefore the change in potential energy?

We have done an experiment to answer this question, using the apparatus shown in Fig. 17.5 and 17.6. Here the descending object is a loaded cart that rolls down the incline. The wheels of the cart have ball

The Change in Gravitational Potential Energy over Different Paths 17.4

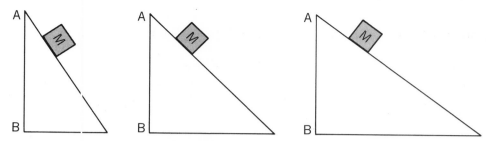

Fig. 17.4 A mass M moving down inclines of different length, but with the change in height AB always the same.

Fig. 17.5 Apparatus for measuring the heat produced when a loaded cart moves slowly down an incline.

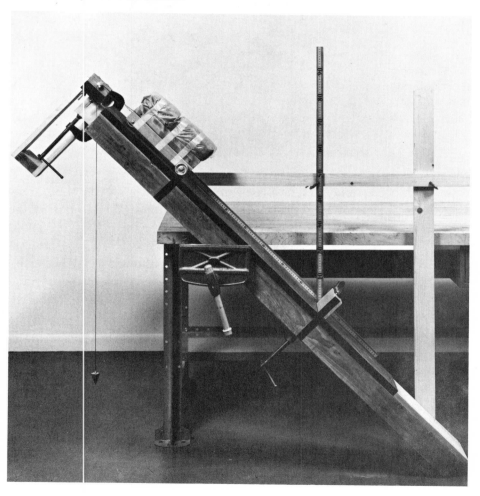

17.4 The Change in Gravitational Potential Energy over Different Paths

Fig. 17.6 A close-up of the top of the apparatus in Fig. 17.5, showing the thread wrapped around the aluminum cylinder and the thermometer inserted in the cylinder.

bearings so that frictional heat losses are kept very low. As the cart slowly rolls down the incline, heat is produced in an aluminum cylinder by the rubbing of a nylon string attached to the cart, just as in Expt. 17.2 and 17.3.

Five runs were made. In one run the cart was allowed to fall vertically like the bucket in Expt. 17.2 and 17.3. The other four runs were made on inclines of different length. In all five runs, the vertical change in height of the cart was the same, 0.50 m. Also, the same counterweight was used in all the runs, the speed of the descent being controlled by changing the number of turns of the nylon string around the cylinder. The results of the experiment are shown in Table 17.1.

As you can see, regardless of the length of the path the cart traveled in changing height by 0.50 m, the temperature change of the cylinder was the same. Therefore, the same quantity of heat was produced in each trial.

We can expect that if we lifted the cart along the inclines by an electric motor instead of lifting it vertically, the "missing heat" would remain the same as long as the vertical rise of the cart remained the same

The Change in Gravitational Potential Energy over Different Paths 17.4

and friction was negligible. We identified the missing heat as an increase in the gravitational potential energy of the cart. Therefore, we can conclude from the measurements in Table 17.1 that the change in gravitational potential energy of an object depends only on the vertical change in its height. It is independent of the sideways change in position.

Table 17.1

Run Number	Length of Path (m)	Vertical Change in Height (m)	Change in Temperature of Aluminum (°C)
1	0.50	0.50	1.8
2	0.58	0.50	1.8
3	0.72	0.50	1.8
4	0.83	0.50	1.8
5	0.97	0.50	1.8

We can generalize our results one step further. From the results described in this section, the loss in gravitational potential energy when an object moves from A to D (Fig. 17.7) is the same as when it moves from A to B. Similarly, the loss in potential energy when it moves from D to C is the same as when it moves from B to C. Hence, whether the object moves along the path ABC or ADC, the same amount of gravitational potential energy is lost.

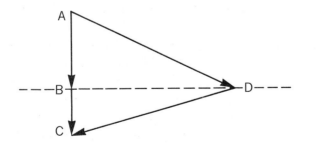

Fig. 17.7 The same amount of gravitational potential energy is lost by an object when it moves from A to C by either path ABC or path ADC.

To sum up, the change in gravitational potential energy depends only on the mass and the change in height. It does not depend on the path.

10† In this experiment the mass of the cart that rolled down the incline was 5.0 kg. What change in temperature of the aluminum cylinder would you predict when the cart descends a vertical distance of 0.30 m?

17.5 Heat Generated by a Contracting Spring

11† In an experiment similar to the one described in Sec. 17.4 the same cart was allowed to descend a series of paths such that the net change in height was always 1.00 m. What changes in temperature of the aluminum cylinder would you expect to find in this experiment?

12 A 5.0-kg mass is lifted vertically from the floor a distance of 2.5 m and then allowed to fall onto a table top that is 1.0 m above the floor. What is its *net* change in gravitational potential energy from beginning to end?

13 What is the *net* change in gravitational potential energy of a body of mass 2 kg, if it moves from A to E along the path shown in Fig. A?

Fig. A For prob. 13.

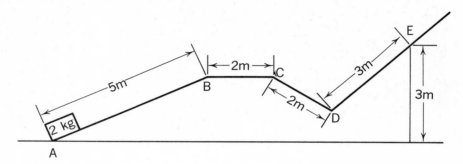

17.5 Heat Generated by a Contracting Spring

Gravitational potential energy is not the only form of potential energy. We can show the existence of another common form by the following demonstration. A spring is held fixed at one end. If the other end is pulled and then suddenly released, the spring will snap back. But if the free end of the stretched spring is attached to a string wrapped around a 14.1-g aluminum cylinder, as seen in Fig. 17.8, the spring will contract slowly, pulling the string with it and generating heat in the cylinder. The amount of heat generated can be determined by the rise in temperature of the cylinder. If we stretch the spring to different initial lengths and let it contract to the same final length, we find different rises in temperature. This is shown in Table 17.2.

If you think of the change in position of the free end of the spring and compare it with the change in position of the falling object in Expt. 17.3, Gravitational Potential Energy as a Function of Height, you can see some similarities in the two experiments. In both cases a change of position of one object is accompanied by an increase in thermal energy in another.

Fig. 17.8 The free end of a stretched spring pulls a nylon string wrapped around an aluminum cylinder and generates heat as the spring contracts.

Table 17.2

Position of Free End of Spring (m)		Change in Position (m)	Rise in Temperature of Cylinder (°C)	Increase in Thermal Energy of Cylinder (joules)
Initial	Final			
0.50	0.20	0.30	0.75	10
0.60	0.20	0.40	1.10	14
0.70	0.20	0.50	1.55	20
0.80	0.20	0.60	2.10	27
0.90	0.20	0.70	2.70	34
1.00	0.20	0.80	3.40	43

Elastic Potential Energy   17.6

An object does not rise by itself. If we get the object to the higher position by lifting it with an electric motor, less heat is generated in the motor than when the motor runs freely for the same electrical work. The "missing heat" is recovered when the object returns to the lower position.

Similarly, a spring does not stretch by itself. If we use an electric motor to stretch it, less thermal energy would be generated in the motor for the same amount of electrical work than if it were running freely. Again, the "missing heat" can be recovered by letting the spring contract as in Fig. 17.8.

We can use a battery to generate a given quantity of heat directly in a motor. The battery can also generate heat indirectly by having the motor first lift an object or stretch a spring. We can then recover the "missing heat." Thus, stretching a spring—that is, changing the position of its free end—changes the energy of the spring. This change of energy is related to the elasticity of the spring and is therefore called *elastic potential energy*.

Changes in gravitational potential energy and elastic potential energy can be converted not only into changes in thermal energy but also into each other. Consider an object hung on a spring but supported by hand [Fig. 17.9(*a*)]. When the object is released, it moves up and down. Fig. 17.9(*b*) shows it in its lowest position. In this position the gravitational potential energy is at a minimum, but the elastic potential energy is at a maximum because the spring is stretched most. In Fig. 17.9(*c*) the object is at the top of its rise. Notice that it is practically at the same position from which it was released [Fig. 17.9(*a*)]. Thus the object regained the gravitational potential energy it lost during the fall, and the spring lost the elastic energy it gained during the fall. The two forms of potential energy are convertible from one to the other.

For Home, Desk, and Lab

14 In a certain run of the apparatus used in Expt. 17.2, the bucket and its contents have a total mass of 1,200 g. The mass of the counterweight is 240 g. What is the net mass that falls during the run?

15 Suppose in Expt. 17.2 that the nylon string broke just as you were starting a run, and the bucket fell to the floor. What do you think happened to the gravitational potential energy that the bucket and its contents lost during the fall?

16 Suppose you measured the electrical work done by an electric motor while it dragged a block of some kind over the surface of a table. Would you expect the electrical work to be equal to the heat produced in the motor? If not, where might you look for the difference?

For Home, Desk, and Lab   107

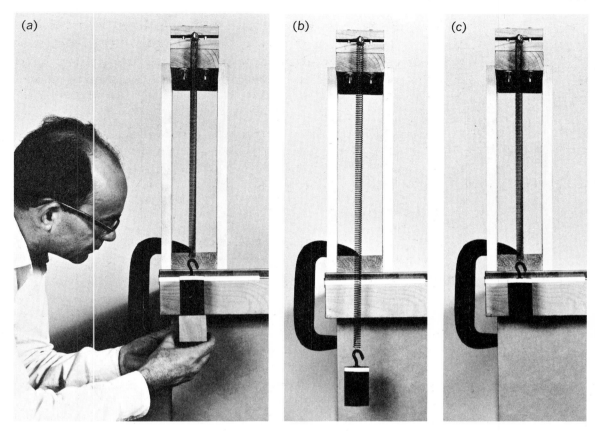

Fig. 17.9 (a) An object hung on the end of a spring is supported by a block of wood held up against the bottom of the table top. Note that the white tape on the top of the object is at the same level as the white tape attached to the table top. (b) When the block of wood supporting the object in (a) is suddenly removed, the object falls and stretches the spring. This photograph shows the object in its lowest position and the spring stretched the maximum amount. (c) The spring then pulls the object up very nearly to its original starting point.

17 How much gravitational potential energy does a 5-kg mass gain if it is lifted:
  a) 1 meter?
  b) 2 meters?
  c) 10 meters?

18 What is the gain in potential energy of the counterweight in Expt. 17.2?

19 When the change in height of an object is expressed in meters and its mass is expressed in kilograms, its change in gravitational potential energy in joules is given by 9.8 × mass × change in height. What are the units of the constant 9.8?

20 After you have climbed up a rope to a height of 5 m, you slide down.
    a) Estimate how much gravitational potential energy you lost coming down.
    b) Where did it go?

21 Ten students whose total mass is 700 kg ride down on an elevator a distance of 30 m (about 10 stories). How much heat is produced in the elevator system by the loss in gravitational potential energy of the students?

22 A tall, narrow cylinder contains 3×10^3 cm^3 of water to a depth of one m. Suppose you let a cylinder of lead having a mass of 1 kg fall through the water starting from rest at the top.
    a) How much heat is generated?
    b) Neglecting the heat absorbed by the lead, what is the rise in temperature of the water?
    c) Could you measure the increase in the temperature of the water with your thermometer?

23 A 1-kg mass loses 1.7 joules of gravitational potential energy in falling 1.00 m on the surface of the moon. If you were to do Expt. 17.2 on the moon, would the temperature rises of the aluminum cylinder be more or less than those you found in the laboratory? How many times more or less?

24 Suppose the frictional heat produced in the wheels of the cart described in the experiment in Sec. 17.4 was so large that it could not be neglected. How do you think this would affect:
    a) the total change in gravitational potential energy of the cart?
    b) the total amount of heat produced as the cart descended along an incline?
    c) the change in temperature of the aluminum cylinder?

25 When an automobile coasts down a hill and is braked to a stop at the bottom, where does the lost gravitational potential energy go?

26 *a)* Use the data in Table 17.2 to draw a graph of the rise in temperature of the aluminum cylinder as a function of the change in position of the free end of the spring.
    b) Is the change in potential energy proportional to the change in position of the free end of the spring?

27 In the experiment described in Sec. 17.5 we did not take into account the fact that when the spring contracted it lifted a 220-g counterweight in addition to heating up the aluminum cylinder.
    a) How does the change in gravitational potential energy of the counterweight compare with the heat produced in the cylinder when the spring contracted from 1.00 m to 0.20 m?
    b) If we were able to do the experiment without a counterweight, what would have been the rise in temperature of the cylinder when the spring contracted from 1.00 m to 0.20 m?

28 A bucket of mass 10 kg is connected to a spring and allowed to fall. The bucket reaches its lowest position after falling 0.40 m. There it is caught before it can go back up. The bucket is removed and the spring is allowed to contract pulling a thread over an aluminum cylinder. About 40 joules of heat is produced. How does this heat compare with the loss in gravitational potential energy of the bucket during the fall?

18 Atomic Potential Energy

At the beginning of Chapter 16 we raised the following question: "Does the relation: Heat (in joules) = electrical work (in joules) hold under all conditions?" This has been answered in part through experiments with electric motors. The results of these experiments led us to realize that when charge flows through a motor that lifts an object or stretches a spring, the heat produced will be less than the electrical work. In both cases the "missing heat" is accompanied by a change in the position of something. In one case the missing heat can be recovered by letting the object fall, in the other by letting the spring contract. These observations led us to the idea of potential energy. Are there other cases where missing heat is related to the change in position of something—cases that may, therefore, suggest other forms of potential energy?

When you boil water with an electric heater, the temperature stays constant while steam is being formed. Electrical work is being done, but there is no change in the thermal energy of the water. Heat is "missing." It is plausible that this missing heat is associated with the change of the water from liquid to gas. In this change the distance between molecules increases.

Can this missing heat be recovered when the vapor is allowed to condense? We shall answer this question in two steps. First we shall find how much electrical work is needed to evaporate 1 g of water at the boiling point. Next we shall measure how much heat is released when 1 g of vapor at the boiling point condenses. Then we shall compare this heat with the electrical work needed to evaporate 1 g.

18.1 The Missing Heat When Water Evaporates

To find out how much electrical work is needed to change 1 g of water at 100°C to water vapor (steam), we used the apparatus shown in Fig.

18.1. Half of the styrofoam insulation has been removed to enable you to see the details. Water in the small flask was heated and boiled by electric charge flowing through a heater in the bottom of the flask. If the heat leakage were negligible, the experiment would be easy to do. When the water was boiling, we would disconnect the apparatus, immediately mass it, and then reconnect it and boil water for a measured length of time, recording both the current and the voltage. Then, after massing again, we could calculate the mass of water evaporated, as well as the electrical work. The electrical work in joules divided by the mass evaporated would give us the value we are after.

However, with the apparatus shown in Fig. 18.1 it was necessary to keep the flask at the relatively high temperature of 100°C for about 200 seconds in order to evaporate enough water to measure its mass accurately. This meant that, even though the flask was insulated, there was enough time for a large amount of heat to leak away through the styrofoam into the surrounding air.

Fig. 18.1 The electrical work supplied to the heater in the flask vaporizes the water. The vapor leaves the flask through the tube leading to the outside. The battery that supplies the electrical work is not shown. The mass of the vapor was determined by the apparatus on the balance just before and after each run. Before timing a run, we boiled the water for a few minutes to make sure that the styrofoam insulation had warmed up as much as it could, so that the rate of heat leakage would remain constant.

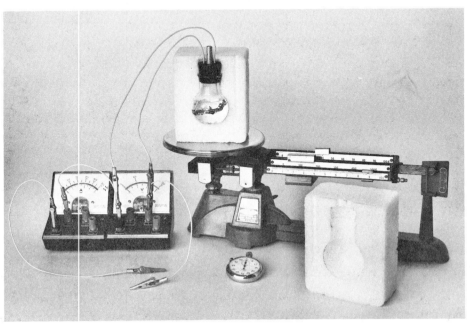

Ideally we would like to do a large amount of work in a very short time, say a few seconds, so that there will be no time for a significant quantity of heat to leak through the insulation. However, the amount of work must be large enough to evaporate a measurable mass of water. The trouble is that if we did this, the boiling would be so vigorous that water would shoot out of the tube in the top of the flask.

What we can do, and did, is to supply the same amount of work (7,000 joules) for shorter and shorter times, measure the mass of water which evaporated in each case, and draw a graph of mass of water evaporated as a function of time. We can then extend the graph to zero time to find the mass of water that would have evaporated if there had been no heat leakage. The data for the experiment are shown in Table 18.1, and the graph is shown in Fig. 18.2. As you can see from the graph, the straight line when extended meets the vertical axis at 3.08 g. We conclude from the graph, therefore, that if there had been no heat leakage, 7,000 joules of electrical work would have vaporized 3.08 g of water at 100°C and that the electrical work needed to vaporize 1 g at 100°C is

$$\frac{7.0 \times 10^3}{3.08} = 2.27 \times 10^3 \text{ joule.}$$

Table 18.1

For a given voltage the current was measured, and the time needed to yield 7.0×10^3 joules of work was calculated. Several runs were made for each voltage chosen. The last column gives the average mass for each run.]

Voltage (volts)	Current (amp)	Time (sec)	Work (joule)	Mass Evaporated (g)
3.5	2.5	816	7.0×10^3	1.35
4.0	2.8	625	7.0×10^3	1.90
4.5	3.2	495	7.0×10^3	2.04
5.0	3.5	400	7.0×10^3	2.22
6.0	4.2	278	7.0×10^3	2.53
7.0	4.9	204	7.0×10^3	2.67

Experiment
18.2 Heat of Condensation of Water

In the preceding section we described an experiment to measure the electrical work needed to vaporize 1 g of water at 100°C. When steam condenses in cold water, the temperature of the water rises. Does an

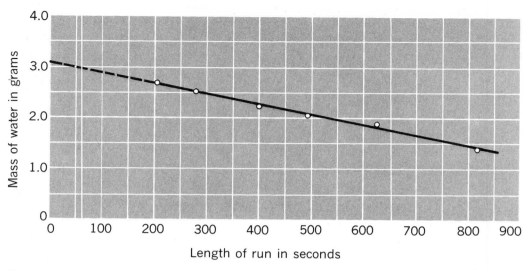

Fig. 18.2 The graph of the mass of water evaporated, using the apparatus shown in Fig. 18.1, as a function of the length of the run.

increase in thermal energy of the water when 1 g of steam condenses equal the electrical work needed to vaporize 1 g of water? You can answer this question in the laboratory.

By condensing some steam in a known mass of water, you can measure the temperature change of the water and from the data calculate the heat of condensation.

The apparatus you will need is shown in Fig. 18.3. In the previous experiment, it was necessary to keep the flask and its contents at 100°C for about 800 sec. This was why we had to design the experiment to allow for a large heat loss. When you condense steam in this experiment,

Fig. 18.3 After boiling chips are added to a test tube about one-third full of water, steam is produced by heating with alcohol burners. The steam can be condensed by placing the glass tube in cold water in the calorimeter.

18.2 Experiment: Heat of Condensation of Water

however, you can reduce the heat loss nearly to zero. You do this by starting with water cooled below room temperature and condensing steam into it until it warms up to a temperature as far above room temperature as it was below room temperature at the start. This is a method you used in Expt. 15.1 when you measured the heat produced in an aluminum cylinder by a flow of electric charge.

After massing the calorimeter cup and cover, add about 50 cm³ of precooled water and remass to find the total mass and the mass of the cold water. Be sure to record the initial temperature of the cold water in the calorimeter just before you start to condense steam.

To prevent water from being sucked back into the test tube, it is advisable to heat the test tube vigorously. You can keep the temperature of the water in the calorimeter fairly uniform while it warms up by gently swirling the water as the steam condenses in it. When the water reaches the temperature you want, disconnect the calorimeter from the steam generator, determine the final temperature of the water, and make a final massing to determine how much steam condensed.

The steam not only condensed to form water at 100°C, but this water at 100°C cooled down to the final temperature of the cool water that you used to condense the steam. Thus the increase in the thermal energy of the cool water comes from two sources, from the condensation of the vapor and from the cooling of the resulting water. How can you calculate each?

How does your value for the heat released when 1 g of steam condenses at 100°C compare with the electrical work needed to vaporize 1 g of water at 100°C?

——— ——— ———

The quantity of heat released when 1 g of vapor condenses is called the *heat of condensation*. The "missing heat" when 1 g of liquid evaporates is called the *heat of vaporization*.

The value of 2.27×10^3 joule/g = 543 cal/g for the heat of vaporization of water that we found in Expt. 18.1 differs by less than 1 percent from 539 cal/g, a value determined by more accurate experimental methods. The heat of vaporization is a characteristic property and is different for different substances. (See Table 18.2.)

1. It takes 2.3×10^3 joules to change 1.0 g of water at 100°C to water vapor at 100°C. How high above the surface of the earth would you have to lift 1.0 g of water to increase its gravitational potential energy by an amount equal to the increase in molecular potential energy due to evaporation?

2† It requires 950 cal to boil away 10 g of sulfur dioxide at its boiling point, which is −10°C. What is the heat of vaporization of this substance?

Table 18.2

Substance	Boiling Point (°C)	Heat of Vaporization at the Boiling Point (cal/g)
Aluminum	2,500	2,500
Ammonia	−33	330
Bromine	59	44
Copper	2,300	1,200
Ethyl alcohol	79	204
Glycol	200	190
Helium	−270	5
Hydrogen	−250	54
Isopropyl alcohol	82	160
Methyl alcohol	65	260
Moth flakes	218	76
Oxygen	−180	51
Zinc	910	420

Atomic Potential Energy   18.3

It is worthwhile to compare the results of the two experiments in the last two sections with the motor experiments you did in Chapter 16. Figure 18.4 illustrates what happens when an electric motor lifts an object. Electrical work is done, the vertical position of an object changes, and there is "missing" heat that can be recovered when the object is allowed to fall slowly to its initial position. Similarly, Fig. 18.5 illustrates that when electrical work is used to vaporize water, there is missing heat and, as you found in the last section, this missing heat can be recovered when the steam is condensed. Unlike the case of a motor lifting an object, when a liquid evaporates, no obvious changes of position take place which we can associate with the missing heat; no visible springs stretch, and the vapor as a whole need not increase its distance from the earth. However, if we look at what happens on the atomic scale when a liquid evaporates, the similarity between lifting an object and vaporizing water becomes apparent.

The density of water vapor is much less than the density of water. The same is true for lead vapor and liquid lead, or for any vapor and its liquid. This means that a given mass of a substance occupies a much larger volume as a vapor or gas than it does as a liquid. Therefore, in the vapor the atoms (or molecules) of a substance are much farther apart

18.3 Atomic Potential Energy

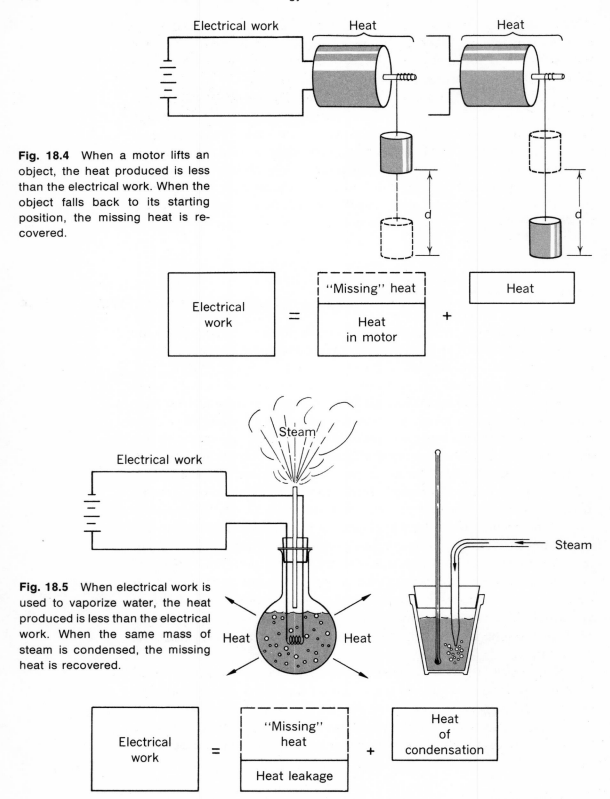

Fig. 18.4 When a motor lifts an object, the heat produced is less than the electrical work. When the object falls back to its starting position, the missing heat is recovered.

Fig. 18.5 When electrical work is used to vaporize water, the heat produced is less than the electrical work. When the same mass of steam is condensed, the missing heat is recovered.

than in the liquid. Experience shows that when atoms (or molecules) are separated from each other in this process, heat is absorbed, and when they come close together again, heat is released. This gives rise to the following extension of the atomic model of matter:

Atoms and molecules have a form of potential energy which depends on the distance between them, just as the gravitational potential energy of an object depends on its distance above the earth. When a liquid evaporates, atoms or whole molecules move away from each other, increasing the atomic or molecular potential energy. When a vapor condenses, the molecules move closer together, the atomic or molecular potential energy decreases, and heat is released, just as heat is released when an object falls or a spring contracts.

Experiment
Heat Produced in the Decomposition of Water   18.4

We have seen that there is a change in atomic potential energy when the atoms or molecules of an element or compound separate or come together in the process of vaporization and condensation.

In these processes all molecules (or atoms, in the case of some elements) either separate or they all come together. In most atomic reactions, in particular in those in which compounds are formed or decomposed, the situation is more complicated. Some atoms get closer to one another, while other atoms get farther apart. In the decomposition of water by electrolysis, for example, a complex rearrangement of atoms takes place: oxygen atoms are separated from hydrogen atoms, but then oxygen atoms come closer together to form oxygen molecules. Hydrogen molecules are formed in a similar way. Does the total atomic potential energy increase or decrease? We can answer this question by comparing the heat produced during the decomposition of water with the electrical work done.

In this experiment you will use a styrofoam calorimeter cup and stainless steel electrodes in a solution of sodium carbonate, as shown in Fig. 18.6. There is no need to collect either the hydrogen or the oxygen produced, since you can determine the masses of these gases from the quantity of charge that flows through the cell.

Support the electrodes by means of a short length of wire unraveled from the electrode and pushed through the lid of the styrofoam calorimeter (Fig. 18.7).

Fill the cup with 30 cm^3 of sodium carbonate solution, place the lid on the top, and insert the thermometer. You can use the circuit shown

18.4 Experiment: Heat Produced in the Decomposition of Water

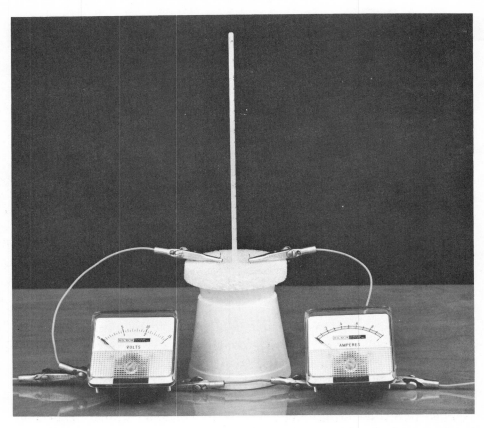

Fig. 18.6 The styrofoam calorimeter used in determining the heat produced when water is decomposed by electrolysis.

Fig. 18.7 The photograph shows how the stainless-steel electrodes are fastened to the lid of the calorimeter. The length of wire at the edge of the wire screen is unraveled halfway. It is then bent into a vertical direction and pushed up through the styrofoam lid. To keep the electrode in the correct position, it can be gently pushed up so that its top edge sticks into the lid.

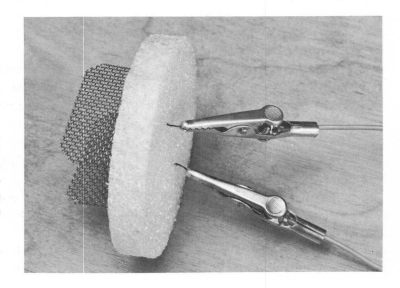

Experiment: Heat Produced in the Decomposition of Water 18.4

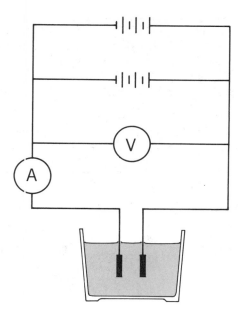

Fig. 18.8 The circuit used in measuring the heat produced in the electrolysis of water. The battery consists of two parallel branches of three cells connected in series.

in Fig. 18.8; but do not connect the cell to the battery until you have measured the initial temperature of the solution in the cell.

After reading the initial temperature, you can connect the sodium carbonate cell to the battery and read the voltage and current every 30 seconds for a few minutes until you observe a temperature change that you can read with reasonable accuracy. Stirring the solution while the charge flows may disturb the connections, so it is best to stir only after shutting off the current before your final temperature reading.

To find the heat released in the solution you must know its heat capacity.

$$\text{Heat capacity} = (\text{specific heat}) \times (\text{mass})$$

Since you are dealing with a liquid, it will be convenient to find the mass indirectly by measuring the volume of the solution and multiplying it by the density. Thus,

$$\text{Heat capacity} = (\text{specific heat}) \times (\text{density}) \times (\text{volume})$$

The product of the first two factors is the heat capacity of one cubic centimeter of the solution. The specific heat of the sodium carbonate solution is slightly less than that of water, and its density is slightly greater. This yields a heat capacity close to 4.2 joules/°C for one cubic centimeter of solution. Use this heat capacity, the volume of the solution, and the temperature rise to find how much heat was released in the cell.

How does the increase in thermal energy of the solution compare with the electrical work?

When hydrogen and oxygen combine to form water, 1.42×10^5 joules of heat are given off for each gram of hydrogen that reacts. Suppose you had collected the hydrogen and oxygen produced in this experiment and then combined the gases to form water. Would the heat produced equal the heat that was missing when you electrolyzed the water?

To answer this question you must first calculate how much hydrogen was produced in your experiment. You know that 1.00 amp-sec passing through the cell produced 1.04×10^{-5} g of hydrogen (Sec. 11.8). From the total charge that flowed through the cell, how much hydrogen was produced? How much heat would be given off if this mass of hydrogen reacted with oxygen? How does this quantity of heat compare with the missing heat in your experiment?

3† Suppose that in this experiment the temperature of the room dropped by two or three degrees while the experiment was being run. (Somebody opened a window, and it was cold outdoors.) How would this affect your results?

4 When you compute the increase in the thermal energy of the sodium carbonate solution in Expt. 18.4, do you have to take into account the thermal energy carried away by the escaping hydrogen and oxygen?

18.5 Chemical Energy and Heats of Reaction

The atomic potential energy involved in the formation of compounds or the breaking up of compounds is often called *chemical energy*. Changes in chemical energy are usually determined by measuring directly the heat produced (or absorbed) during reactions. Since these heats of reaction are found by calorimetric measurements, their values are given in tables in cal/g rather than in joules/g. The heats of reaction of some common reactions are given in Table 18.3. When 1 g of nitrogen combines with oxygen to form nitrogen oxide, 1.5×10^3 cal are *absorbed* in the formation. To indicate this, the heat of reaction in this case is given a negative sign. In many reactions the heat of reaction goes unnoticed either because it is small or because the reaction takes place so slowly (the rusting of iron, for example) that the heat is lost to (or gained from) the surroundings nearly as fast as it is released (or absorbed).

Table 18.3

Substance	Reaction	Heat of Reaction Per Gram of Substance (cal/g)
Methanol	Methanol + oxygen ⟶ carbon dioxide + water	5.3×10^3
Gasoline	Gasoline + oxygen ⟶ carbon dioxide + water	11.5×10^3
Furnace oil	Furnace oil + oxygen ⟶ carbon dioxide + water	11.0×10^3
Hydrogen	Hydrogen + oxygen ⟶ water	33.9×10^3
Nitrogen	Nitrogen + oxygen ⟶ nitrogen oxide	-1.5×10^3
Magnesium	Magnesium + hydrochloric acid ⟶ hydrogen + magnesium chloride	1.2×10^3
Sodium	Sodium + chlorine ⟶ sodium chloride	4.3×10^3
Copper	Copper + sulfur ⟶ copper sulfide	0.18×10^3
TNT (trinitrotoluene)	TNT ⟶ nitrogen + hydrogen + water + carbon monoxide + carbon dioxide	less than 3.6×10^3

5 Sodium reacts spontaneously with water. The graph in Fig. A. shows the amount of heat produced when different amounts of sodium are added to 100 cm³ of water. What is the heat of reaction of sodium?

6 Does Table 18.3 indicate why liquid hydrogen is one of the best fuels for propelling rockets that carry satellites into orbit?

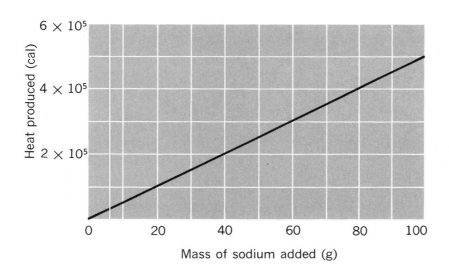

Fig. A For prob. 5.

For Home, Desk, and Lab

7 If the straight-line graph in Fig. 18.2 is extended to the right, it meets the horizontal axis at time = 1,500 sec. What does this tell you about what happens when you make a 1,500-sec run? Be as complete as you can in your description.

8 From the graph in Fig. 8.2, and knowing that 7.0×10^3 joules of work was used in each run, determine
 a) the heat lost to the surroundings in a 500-sec run.
 b) the rate (in joules/sec) at which heat leaked out of the apparatus.

9 Suppose that in the experiment using the apparatus in Fig. 18.1 the work was 3,500 joules instead of 7,000 joules. How would this affect the graph in Fig. 18.2?

10 If you moisten the back of your hand, it soon dries off; but while it is drying, your hand feels cool. How do you explain this?

11 Why is it that a burn from steam at 100°C is more severe than one from air at 100°C?

12 In calculating the increase in the thermal energy of the calorimeter in Expt. 18.4, we took only the solution into account. Estimate the error introduced by not considering the electrodes.

13 Suppose when you measured the heat produced in the decomposition of water (Expt. 18.4) you used a 4-volt battery instead of a 3-volt battery. For a given flow of charge, how would this affect
 a) the mass of hydrogen produced?
 b) the electrical work done?
 c) the missing heat?
 d) the change in thermal energy of the solution?

14 How would you expect the heat produced when 1.0 g of copper dissolves in a copper-plating cell to compare with the heat produced when 1.0 g of copper plates out of the solution?

15 Suppose some astronauts landed on a planet whose atmosphere was composed of hydrogen.
 a) What could they use for fuel in such an atmosphere?
 b) How much heat per gram of fuel would they obtain? (See Table 18.3.)

16 Recall the cooling-curve graph you got when you found the freezing point of moth flakes. In the light of what you have learned in this chapter, how do you explain the flat section of the graph (the plateau) that represented what happened while the moth flakes were freezing?

17 The heat required to melt 1 g of ice at the melting point is called the heat of fusion. Design an experiment to measure the heat of fusion.

Kinetic Energy 19

Experiment
Heat Generated by a Rotating Wheel   19.1

Suppose we connect a mass to a string that is wrapped around the hub of a bicycle wheel as shown in Fig. 19.1. As the mass falls, it turns the wheel; if the wheel bearings are good, there is very little friction, and very little heat is generated in them. Since the string does not slide around

Fig. 19.1 When the bicycle wheel is released, the bucket containing bags of sand unwinds the string on the hub, setting the wheel in motion. At right is a close-up of the hub, showing how a loop on the end of the string is hooked over a screw on the drum. This allows the string to be released automatically from the drum when it is unwound.

19.1 Experiment: Heat Generated by a Rotating Wheel

the hub but merely unwinds, no heat is generated by the string as the mass falls. Instead, the decrease in gravitational potential energy of the falling mass is accompanied by an increase in the speed of the rotating wheel.

What has happened to the gravitational potential energy lost by the falling mass?

Perhaps it has been changed into a form of energy associated with the motion of the wheel. If this is the case, the spinning wheel should be able to generate heat in coming to a stop. We can, of course, bring the wheel to a stop by having it rub on something. Rubbing will, as you know from previous experiments, produce heat. Suppose a wheel is set in motion by a falling mass, and the wheel is then stopped by having it rub on something. How will the heat produced in stopping the wheel compare with the heat produced by the same mass falling slowly while pulling a string around a metal cylinder as in Expt. 17.2?

To find out, we can let a falling mass turn the wheel as it descends. As soon as the mass reaches the floor, we can have the string automatically released from the wheel. We can then hold an aluminum cylinder containing a thermometer against the spinning hub and measure the temperature rise in the aluminum cylinder as the wheel is brought to rest. Figure 19.2 shows how this can be done.

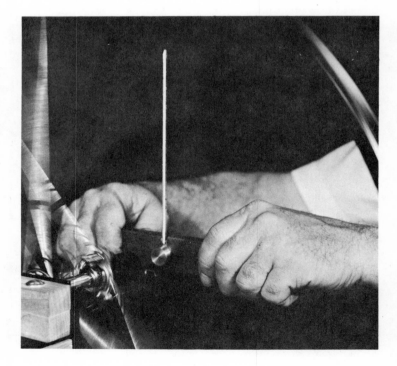

Fig. 19.2 The rotating wheel can be stopped by friction between the rotating hub and an aluminum cylinder pressed against it. A thermometer inserted in the cylinder measures its temperature rise.

Support a mass of 2 or 3 kg from the hub of a bicycle wheel as shown in Fig. 19.1. Allow the mass to fall the same distance as in Expt. 17.2; then stop the spinning wheel as shown in Fig. 19.2 and measure the temperature rise of the aluminum cylinder.

How does the heat produced in this experiment compare with the heat produced when the same mass fell slowly through the same distance in Expt. 17.2? (The heat capacity of the aluminum cylinder is 13.4 joules/°C.)

1   What additional causes of heat loss occur in Expt. 19.1 as compared with Expt. 17.2 and 17.3?

2   If the friction in the bearings of the wheel used in Expt. 19.1 increases as the wheel wears through use, what would you expect to happen to the temperature changes you record for the aluminum cylinder?

Another Form of Energy   19.2

You have now observed a mass falling slowly under two different conditions. First the string holding the mass was rubbing against an aluminum cylinder. The heat generated in the cylinder was about the same as the heat that would be "missing" if the mass had been lifted the same distance by an electric motor. In the second case, when the falling mass turned the wheel, the heat was still missing after the mass fell. However, you found that the missing heat was produced in stopping the wheel. To put it differently, heat can be produced when a body loses height or loses speed. Thus, from these two experiments we see that a loss in height of a body is equivalent to a loss of speed. When a body loses height, it loses gravitational potential energy. We conclude, therefore, that when a body loses speed, it loses another form of energy. This is called *kinetic energy*. When an object is put in motion, it gains kinetic energy, just as any object that is lifted gains gravitational potential energy.

We saw in Sec. 17.6 that gravitational potential energy could be changed into elastic potential energy, which in turn could be changed back into gravitational potential energy. Can either of these forms of energy be converted into kinetic energy and back again? Suppose a falling mass turns a wheel; will the turning wheel be able to lift the mass back to its original position?

19.2 Another Form of Energy

Fig. 19.3 Mass at (a) starting position; (b) lowest position; (c) final position.

Figure 19.3 shows what happens when the experiment is done. The apparatus is similar to the one you used in your last experiment, except that the end of the string is fastened to the hub of the wheel so that it will not fall off when it is unwound.

Note that the mass comes back up very close to its original position, but the final position is a little below the initial one. Thus, almost all the initial gravitational potential energy was converted to kinetic energy and back. Is the rest lost? Although the wheel turns very smoothly, there is a little friction in the wheel bearings and a little friction between the moving wheel and the surrounding air. It is very hard to measure the increase in thermal energy of the bearings and the air during one fall and rise of the mass. However, if an experiment like this were done in an evacuated chamber and with better bearings, the friction would be reduced and the mass would rise even closer to its starting point. We may generalize from such experiments that if we could eliminate friction entirely, the whole process shown in Fig. 19.3 could take place without any rise in temperature, and the final position of the mass would be the same as the initial one.

Figure 19.4 shows a similar experiment in which the falling mass is replaced by a contracting spring. Between (*a*) and (*b*) the elastic potential energy of the spring decreased, and the kinetic energy of the wheel increased. Between (*b*) and (*c*) it is the other way around. Finally, in (*c*) the wheel is again at rest and the spring is stretched almost as much as in (*a*).

3† Suppose that a 5-kg mass is substituted for the 1-kg mass shown in Fig. 19.3. Will it rise as high as the 1-kg mass does?

Fig. 19.4 (*a*) A stretched spring is connected to the string wound around the wheel hub. (*b*) When the string is completely unwound, the spring has contracted to its shortest length and the wheel has reached its maximum speed. (*c*) The spinning wheel has stretched the spring to nearly its original length.

19.3 Kinetic Energy as a Function of Mass

What if we do Expt. 19.1 (Heat Generated by a Rotating Wheel), using a wheel with a different mass? Would we find that the kinetic energy of a moving object depends on its mass? We do not have to do the experiment to find out. Just think of simultaneously doing the experiment with two identical sets of apparatus. The two wheels are put into motion by identical masses falling through the same distance. The final speed of the two wheels would be the same even though they are not connected together, and each would produce the same quantity of heat when stopped separately by friction. The total quantity of heat produced, and therefore the kinetic energy lost in stopping both wheels, would be just twice that of a single wheel. Two wheels have twice the mass of one. If the two wheels were connected together, we would expect the same result. This suggests that for a given speed, if we double the mass of a moving object, we double the kinetic energy. Or, more generally, the kinetic energy of an object is proportional to its mass. This relationship can be verified experimentally by changing the mass of the rim.

Experiment
19.4 Kinetic Energy as a Function of Speed

You have already seen that the kinetic energy of an object depends on its speed, but we have not yet determined the exact relationship between these two quantities. We can find this relationship by measuring the speed of the wheel used in the last experiment when it is put into motion by a mass that falls through different distances. The falling mass loses different amounts of gravitational potential energy, which are gained by the wheel, giving it different speeds.

The rim of the wheel in this experiment has been made very heavy by replacing the tire with a ring of steel. Now almost all the mass of the wheel can be considered to be at practically the same distance from the center of the wheel. Each part of the heavy rim moves, in one revolution, a distance very nearly equal to the circumference of the wheel; the speed of the rim is, therefore,

$$\text{Speed of rim} = \frac{\text{circumference of wheel}}{\text{time for one revolution}}$$

$$= \frac{(2\pi) \times (\text{radius of wheel})}{\text{time for one revolution}}$$

When you measure the radius of the wheel, should you measure it from the center to the inside of the rim, to the outside of the steel rim, or to some other point?

Each time, after the mass reaches the floor, you can use a stop watch to find the time it takes the wheel to make several revolutions. From these data you can calculate, for each run, the time it takes the wheel to make one revolution.

Under what conditions do you think that timing several revolutions will give a more accurate result for the speed of the wheel than timing just one revolution?

Allow a 1-kg mass to fall different measured distances, while it puts the wheel in motion. As soon as the string falls off the drum, time several revolutions of the wheel.

Plot a graph of the speed in m/sec of the rim as a function of the gain in kinetic energy of the wheel.

From the graph, what happens to the kinetic energy when the speed is doubled? When you triple the speed? What do you conclude about the relation between kinetic energy and speed?

4† A toy train goes around its circular track once every 20 seconds. If the circumference of the track is 2.0 meters, how fast is the train going?

5† If a cylinder heats up by 2.8°C while stopping a spinning wheel, how much will it heat up if the rim speed is halved? If it is doubled?

6† Two sets of apparatus are constructed like that used in Expt. 19.1, identical in all respects except that the mass of one of the wheels is twice that of the other. Each of these wheels is set in motion by a 4-kg mass falling through one meter.
a) What do you expect to be the ratio of the quantities of heat generated by the two wheels as they are brought to rest?
b) How do you expect the speeds of the wheels to compare at the instant the cords supporting the 4-kg masses slip free from the hubs?

7 Consider the wheels in Problem 6. If the speed of the lighter wheel is twice that of the heavier one, what will be the ratio of their kinetic energies?

Kinetic Energy, Mass, and Speed 19.5

In the preceding experiment you have investigated the relation between kinetic energy and speed for a rotating wheel with almost all its mass in the rim. The kinetic energy of the wheel was supplied by a mass which lost gravitational potential energy while falling slowly. Will the same relationship hold also under other circumstances? What will be the relationship between kinetic energy and speed for a body sliding in a straight line along a smooth horizontal surface? It turns out that it makes no difference whether an object is moving along a curved path or a straight line. Regardless of the path, the kinetic energy of an object depends only on its mass and speed; it is proportional to the product of its mass and the square of its speed. In other words, the kinetic energy is equal to a constant times the product of mass and the square of the speed:

$$\text{Kinetic energy} = (\text{constant}) \times (\text{mass}) \times (\text{speed})^2$$

You can calculate an approximate value for the constant from the mass of the wheel's rim and the results you obtained in Expt. 19.4, Kinetic Energy as a Function of Speed.

$$\text{Constant} = \frac{\text{kinetic energy}}{(\text{mass}) \times (\text{speed})^2}$$

This constant is the same for all moving objects, so it is important to determine it as accurately as we can. To do this we must consider each of the quantities in the equation above.

Is the kinetic energy gained by the wheel really equal to the gravitational potential energy lost by the slowly falling mass?

Is the speed found by timing 5 to 10 revolutions really the speed the wheel had during the first revolution after the mass reached the floor?

Since not all the mass of the wheel is in the rim, and therefore some of it is moving at a lower speed than the rim, what value for the mass should we use in the calculation of the constant?

Let us consider the kinetic energy first. In Expt. 19.4 you assumed that the kinetic energy of the wheel was equal to the loss in gravitational potential energy of the falling weight. But, as Fig. 19.3 shows, this is not quite true. In Fig. 19.3 we see that the weight does not rise quite as far as it falls, and this means that not all the gravitational potential energy is being converted to kinetic energy and back again.

To find out how much gravitational potential energy was not changed into kinetic energy when the mass fell in Expt. 19.4, we did the experiment shown in Fig. 19.3, measuring the distance a 1.00-kg mass fell and the distance it rose. In five trials the mass was allowed to fall 1.000 m. We found that the height it rose after falling varied from 0.923 m to 0.930 m and the average rise was 0.928 m. This means that only 92.8 percent of the gravitational potential energy was recovered when the mass fell and rose. In other words, $100 - 92.8 = 7.2$ percent of the gravitational potential energy was not recovered when the mass fell and rose again.

(This energy appeared as thermal energy in the bearings and in the air.) In Expt. 19.4 we were concerned only with what happened during the fall, since the mass became disconnected from the wheel when it reached the floor. During the fall alone, one half of 7.2 percent, or 3.6 percent, of the gravitational potential energy was not converted into kinetic energy of the wheel. Therefore, when the wheel is put into motion by a 1-kg mass falling 1 m, its gain in kinetic energy is only $100 - 3.6 = 96.4$ percent of the loss in gravitational potential energy of the falling mass.

Let us now proceed to the examination of the speed of the wheel. If we allow the wheel to spin for a long time, it will eventually come to a stop. In other words, the spinning wheel is continually slowing down. Therefore, the calculated speed of the rim depends on the number of revolutions timed. Figure 19.5 shows the speed calculated from timing different numbers of revolutions. From the graph we can find the speed of the rim during the first revolution, before it slowed down appreciably. As you can see, the speed of the rim during the first revolution was 1.97 m/sec, about 1 percent higher than the speed measured by timing 10 revolutions.

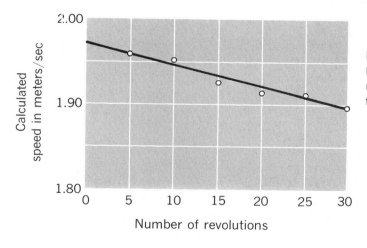

Fig. 19.5 The rim speed of the bicycle wheel as a function of the number of revolutions of the wheel that were timed.

In Expt. 19.4 you assumed that all the moving mass was on the rim, moving at the speed of the rim. In fact, both the spokes and the hub are also moving. Let us look at the contribution of the motion of the hub and spokes to the kinetic energy of the wheel.

The mass of the hub is so close to the center that it is moving very slowly—so slowly that the square of its speed is negligible compared with the square of the rim speed. Therefore, we can neglect its contribution to the total kinetic energy.

19.5 Kinetic Energy, Mass, and Speed

But we cannot neglect the mass of the spokes. Although their inner ends are moving no faster than the outer part of the hub, they extend all the way to the rim, where the speed is high. If we neglect their mass in calculating the constant, we shall be dividing by too small a mass. This will result in too large a value for the constant. On the other hand, if we add the mass of the spokes to the mass of the rim, we shall get a value for the constant that is too low, because the inner parts of the spokes are moving considerably slower than the mass on the rim. However, if we calculate the ratio both ways, first without including the mass of the spokes and a second time including them, we will "bracket" the value of the constant; we know that it lies between the two values we calculate.

We have taken the wheel we used apart and have found the mass of the spokes and rim. Table 19.1 shows the results of these measurements and the values of the constant we found when we calculated it twice, both including and not including the spokes in the calculation.

Table 19.1

Falling mass	1.00 kg
Distance of fall	1.00 m
Kinetic energy of wheel	
0.964×9.8 joules/kg-m $\times 1.00$ kg $\times 1.00$ m $= 9.45$ joules	
Mass of rim	4.78 kg
Mass of rim + spokes	5.03 kg
Speed of rim (from Fig. 19.5)	1.97 m/sec

$$\frac{\text{Kinetic energy}}{(\text{mass of rim}) \times (\text{speed})^2} = \frac{9.45}{4.78 \times (1.97)^2} = 0.51 \frac{\text{joules}}{\text{kg} \times (\text{m/sec})^2}$$

$$\frac{\text{Kinetic energy}}{(\text{mass of rim + spokes}) \times (\text{speed})^2} = \frac{9.45}{5.03 \times (1.97)^2}$$

$$= 0.48 \frac{\text{joules}}{\text{kg} \times (\text{m/sec})^2}$$

The correct value of the constant lies between the two values we have found. Many experiments, different from Expt. 19.4, show that the constant is $\frac{1}{2}$ to a very high degree of accuracy.

8 The radius of the outside of the hub of the bicycle wheel used in the experiment described in this section is 1.4 cm.
    a) What is the ratio of the speed of the outside of the hub to the speed of the steel rim?
    b) What is the square of this ratio?

9† A 1,000-kg automobile is moving at 30 m/sec (about 67 mi/hr) along a highway.
  a) What is its kinetic energy?
  b) If the same automobile were pushed off the edge of a cliff, it would lose gravitational potential energy. About how far (in meters) would it have to fall to arrive at a speed of 30 m/sec? About how many feet is this?

10 A 60-kg boy hangs by his hands from the branch of a tree that overhangs a swimming pool. His feet are 4.0 m above the pool.
  a) If he lets go of the branch, what is his change in gravitational potential energy in falling to the point where his feet just hit the water?
  b) How much kinetic energy does he have at this point?
  c) What is his speed at this point?

A Re-examination of Experiments with Falling Masses 19.6

You have done two experiments, Expt. 19.1 and 19.4, in which a falling mass loses gravitational potential energy while putting a wheel into motion and increasing the wheel's kinetic energy. In each case you assumed that the entire loss in gravitational potential energy of the mass was converted into the kinetic energy of the wheel, while in fact some of it went into the kinetic energy gained by the falling mass. Were you justified in neglecting this kinetic energy? To find out, we can calculate this kinetic energy and compare it with the loss in gravitational potential energy of the descending mass.

To determine the kinetic energy of the falling mass, we need to know its speed. You did not measure its speed in either of the two experiments with the bicycle wheel, so we must find some way of calculating it from the data taken in the experiments. This we can do from the data for Expt. 19.4, Kinetic Energy as a Function of Speed.

In the experiment you measured the speed of the rim, which is directly related to the speed of the mass when it hits the ground. Consider Fig. 19.6. When the wheel rotates through 1 revolution, a point on the rim moves a distance equal to $\pi \times$ (diameter of the wheel).

During the same time, the length of string that unwinds from the hub is $\pi \times$ (diameter of the hub), and the mass descends this distance.

The ratio of these two distances is

$$\frac{\pi \times \text{(diameter of hub)}}{\pi \times \text{(diameter of wheel)}} = \frac{\text{diameter of hub}}{\text{diameter of wheel}}$$

19.6 A Re-examination of Experiments with Falling Masses

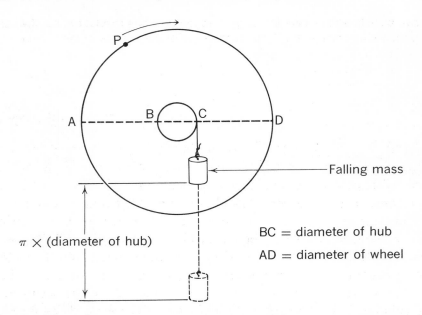

Fig. 19.6 For one revolution of the wheel, any point P on the rim moves a distance π × (diameter of wheel), and the falling weight descends a distance π × (diameter of hub).

and this is also the ratio of the speed of the descending mass to the rim speed, since both distances are traveled in the same time, the time for one revolution of the wheel.

In the apparatus shown in Fig. 19.1, the diameter of the hub is 2.8 cm and the diameter of the wheel is 60 cm, so the ratio of the speeds is

$$\frac{\text{Speed of hub}}{\text{speed of rim}} = \frac{\text{diameter of hub}}{\text{diameter of wheel}} = \frac{2.8}{60} = 0.047$$

and the speed of the hub is 0.047 times the rim speed.

In Table 19.1, the measured rim speed was 2.0 m/sec, so the speed of the descending 1.0-kg mass when it reached the floor was 0.047 × 2.0 m/sec = 0.094 m/sec. The kinetic energy of the falling mass was $\frac{1.0}{2} \times (0.094)^2 = 4.4 \times 10^{-3}$ joule, which is only $\frac{4.4 \times 10^{-3} \text{ joule}}{10 \text{ joules}} \approx 5 \times 10^{-4}$ or about five one-hundredths of 1 percent of the loss in gravitational energy. Obviously, you were justified in neglecting the gain in kinetic energy of the falling mass in Expt. 19.4.

11† Two bicycle wheels take the same time to make one revolution, but the ratio of their diameters is 3/1. What is the ratio of the speeds of their rims?

Thermal Energy of a Gas 19.7

In Chapter 10 you saw a demonstration in which you observed bromine gas diffusing throughout two glass tubes, one containing bromine and air and the other containing only bromine. The observations you made led to an extension of our atomic model. We concluded that in addition to being far apart compared with their size, gas molecules are in rapid motion, moving about in an irregular fashion, occasionally colliding with one another and with the walls of the container that holds them. We further observed in experiments with a "sphere gas" machine that when we increased the speed of the driving motor, the sphere gas occupied a larger volume. This effect suggests that when a gas expands as a result of an increase in temperature, its molecules speed up.

Let us now see how this behavior of gas molecules relates to what we have learned about kinetic and thermal energy. It is clear that moving gas molecules possess kinetic energy, the kinetic energy of each molecule being equal to $\frac{1}{2}$ times its mass times the square of its speed. An increase in speed of the molecules means an increase in their kinetic energies. We already know that an increase in molecular speed means a rise in the temperature of a gas. We can conclude, therefore, that an increase in temperature corresponds to an increase in the kinetic energy of gas molecules.

But an increase in the temperature of a gas means an increase in its thermal energy. Thus we further conclude that the thermal energy of a gas is just the total kinetic energy of its molecules.

A "Disc Gas" Machine 19.8

As you know, the molecules of different gases have different masses. The mass of a molecule of carbon dioxide, for example, is 44 amu compared with a mass of 2 amu for a hydrogen molecule. Consider two gases such as carbon dioxide and hydrogen at the same temperature; how do the speeds of their molecules compare? Do the heavier carbon dioxide molecules move, on the average, at the same speed as the lighter hydrogen

19.8 A "Disc Gas" Machine

molecules, or do they move faster? Or slower? In Chapter 10 we used a "sphere gas" machine to simulate the effects of a change in pressure or temperature on the volume of a gas. We can use a similar machine to suggest an answer to this question.

Figure 19.7 shows a "disc gas" machine that can be used to simulate molecular motion. The discs on top of the shallow box represent gas molecules. When the rim surrounding the top of the box is vibrated rapidly, it keeps the discs in motion, just as the movable platform in the "sphere gas" machine described in Chapter 10 kept small steel spheres in motion in a cylinder. There are two kinds of discs on the top of the box, heavy ones and light ones, to represent heavy molecules and light

Fig. 19.7 A "disc gas" machine that can be used to simulate molecular motion. Compressed air from a vacuum cleaner is blown into the shallow square box through the flexible tube at the right and escapes through many tiny holes in the top surface. The disc "molecules" are supported by the escaping air so that they do not touch the surface of the box: They ride on a cushion of air, with practically no friction. The narrow rim around the top of the box is supported by four short springs, one at each corner, and is connected at the left-hand front corner to a short metal strip which is agitated by an electric motor. When the motor runs, vibrating the rim, the discs, connected to the rim by short lengths of stiff wire, also vibrate. These vibrating discs, which simulate the atoms of the wall of a container of gas, strike the freely moving disc molecules, keeping them in motion.

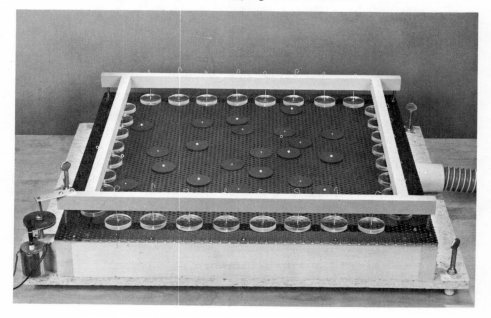

molecules. The heavy discs are marked with a large white dot, and the light ones with a small white dot. When the apparatus has been running for some time, we can assume that enough collisions have taken place so that both "gases" are at the same temperature.

When the discs are in motion, we can measure their speeds from a time-exposure photograph. Figure 19.8 shows a time exposure lasting one-fourth of a second. The white streaks show how far each disc moved while the camera shutter was open. The longest streaks are made by the discs that are moving most rapidly. The short streaks indicate slowly moving discs. A sharp bend in a streak means that during the time exposure, the disc collided with another and bounced off in a new direction. As you can see, the discs are moving in many different directions with different speeds, and occasional collisions occur between discs. On the average, the light discs go farther during the time exposure than do the heavy discs.

To compare the average speed of the heavy and light discs, we measured the lengths of the streaks, which are proportional to the speeds, and calculated the average length for both kinds of discs. We found from Fig. 19.8 that the average speed of the light discs was about twice the average speed of the heavy discs. But we cannot use the evidence from one such photograph alone to come to a conclusion about the ratio of

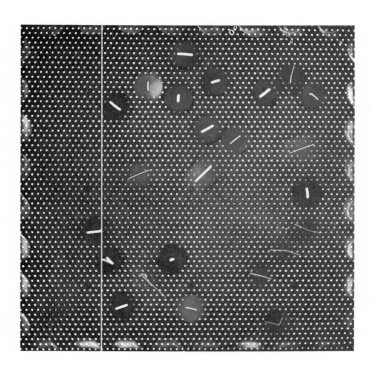

Fig. 19.8 A time exposure of the moving heavy and light discs. In this photograph, the average speed of the light discs is greater than that of the heavy discs.

the average speeds of the two kinds of discs. We have such a small number of discs whose speeds we can measure that it may be just by chance that the average speed of the light discs is greater than that of the heavy discs. In fact, this is borne out by Fig. 19.9, a photograph taken a short time after Fig. 19.8. Here the heavy discs are moving faster, on the average, than the light discs. The ratio of the average speeds of the light to those of the heavy discs is now about 0.9.

If we try to come to a conclusion about the speeds of the discs from one photograph alone, we are in a position corresponding to tossing a penny only a small number of times to find out how often it will turn up "heads." If we toss the penny just six times, say, there is a good chance that only two of the six tosses will come up heads, and there is an equally good chance that four of the six tosses will come up heads. You know, however, that if you toss a penny many times, the ratio of heads to total number of tosses will be close to 1/2, not 1/3 or 2/3. So, if we wish to have confidence in our conclusions about the average speeds of the discs, we must measure the streaks on many photographs of the moving discs. We have done this for a total of 14 photographs. Figure 19.10 shows histograms of the results, which clearly indicate that the light discs move faster, on the average, than the heavy ones.

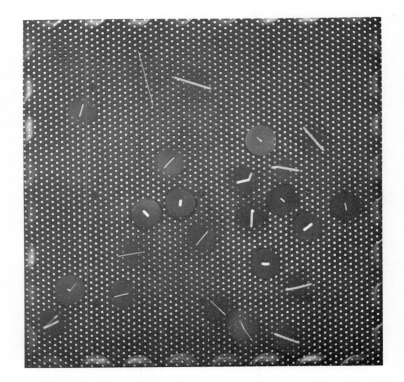

Fig. 19.9 Another photograph of the discs taken at a later time when the heavy discs happened to be moving faster, on the average, than the light discs.

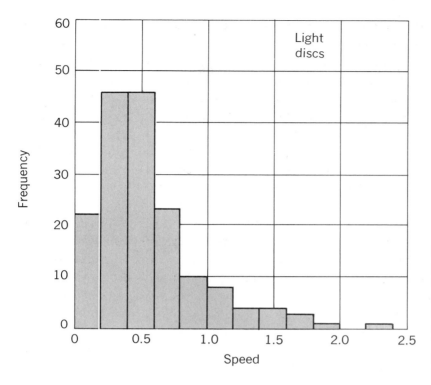

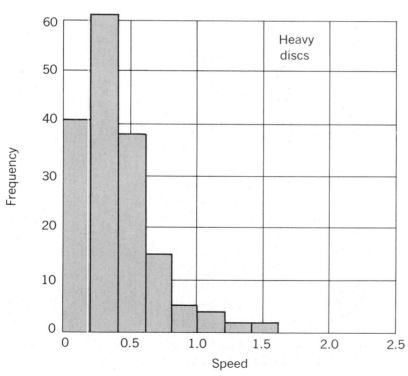

Fig. 19.10 The upper histogram shows the distribution of speeds of the light discs and the lower histogram shows the distribution of speeds of the heavy discs. The average speed of the light discs was 0.57 units and that of the heavy discs was 0.40 units. Both histograms were made from measurements on 14 time exposures, including those shown in Fig. 19.8 and Fig. 19.9, but enlarged to make it easier to measure the lengths of the streaks. The speed units are centimeters measured on the enlarged photographs per fourth of a second. The mass of the heavy discs was almost twice that of the light discs.

If we repeated the experiment and again measured the speeds from 14 more photographs, we would get similar results. Although the histograms would differ slightly in detail, the average speeds would be within a few percent of those shown in Fig. 19.10, because we have measured the speeds of a large number of discs. Such a second run of the experiment would correspond to tossing the penny for a second run of 100 trials. Because of the large number of tosses, we would expect to get close to 50 percent heads in both runs.

Our experiment with the "disc gas" machine leads us to predict that when two gases are at the same temperature, light molecules move faster on the average than heavy ones.

12 Can the speed of a disc in the "disc gas" machine ever change without its colliding with the vibrating discs attached to the rim?

13 In terms of molecular diameters, the molecules of a gas are about 10 diameters apart, on the average, at room temperature and atmospheric pressure. About how far apart, on the average, in terms of disc diameters, are the discs shown in Fig. 19.7 and 19.8?

14† From the results of the "disc gas" experiments discussed in Sec. 19.8, how do you predict that the average speed of the bromine molecules (Br_2, mass 160 amu) compares with the average speed of the molecules of air in the closed tubes described in Sec. 10.2, Molecular Motion and Diffusion?

15 What happens to the total kinetic energy of a gas
    a) if you double the number of molecules?
    b) if you double the speed of the molecules?

Experiment
19.9 The Effusion of Different Gases

How can we test the prediction for gases which was made at the end of the preceding section? Before we answer this question, let us first consider a simpler situation. Suppose you want to find out which of two boys runs faster. You could let both of them run the same distance and time them. The runner who needs less time to cover the same distance runs faster.

We can use a similar method to compare the average speeds of the molecules of two different gases. However, instead of timing one molecule, we shall time a large but equal number of molecules, and instead of timing their motion along a given track, we shall measure the time it takes them to leak out from a container of given volume through a very small hole.

Experiment: The Effusion of Different Gases 19.9

The process of leaking through a small hole is called effusion. In this experiment, you will compare the effusion time of an equal number of molecules of different mass contained in equal volumes and at equal temperatures.

There is no need to count molecules to be sure that we have equal numbers. We can guarantee this by having equal volumes of gas at the same temperature and pressure. To see that this is indeed the case, we calculate the number of molecules in one cubic centimeter as follows:

$$\text{Number of molecules} = \frac{\text{mass of gas}}{\text{mass of one molecule}}$$

$$\text{Number of molecules per cm}^3 = \frac{\frac{\text{mass of gas}}{\text{mass of one molecule}}}{\text{volume of gas}}$$

$$= \frac{\text{mass of gas/volume of gas}}{\text{mass of one molecule}}$$

$$= \frac{\text{density of gas}}{\text{mass of one molecule}}$$

Table 19.2 shows the number of molecules per cubic centimeter for the gases you will be using in this experiment, calculated from their densities at atmospheric pressure and room temperature. The densities are taken from Table 3.1, and molecular masses are calculated from the atomic masses in Table 12.1 and the molecular formulas. It is quite clear from the table that equal volumes of these gases at the same temperature and pressure contain equal numbers of molecules.

A way to compare the effusion times of the different gases is shown in Fig. 19.11. The small balloon is filled with a gas, and the time it takes the gas to escape through the tiny pinhole is measured.

There are several things you must do to make sure that the same volume of gas effuses through the pinhole in each run. To begin with,

Table 19.2

Gas	Density (g/cm³)	Molecular Mass (g)	Number of Molecules per cm³
Air (mostly nitrogen, N_2)	1.2×10^{-3}	4.7×10^{-23}	2.5×10^{19}
Carbon dioxide (CO_2)	1.8×10^{-3}	7.3×10^{-23}	2.5×10^{19}
Hydrogen (H_2)	8.4×10^{-5}	3.3×10^{-24}	2.5×10^{19}

142 19.9 Experiment: The Effusion of Different Gases

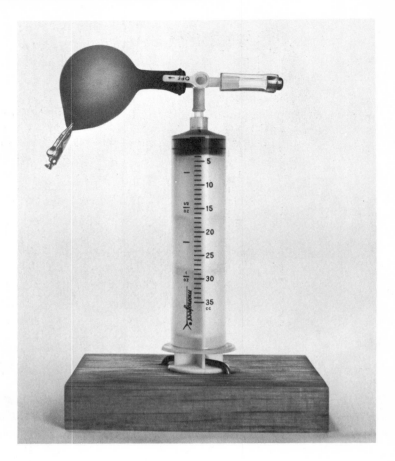

Fig. 19.11 Apparatus for measuring the effusion time of a gas. The handle of the three-way valve always points to the opening that is closed. In the position shown, gas cannot escape from the balloon.

When the valve handle is turned 90° counterclockwise, gas can escape from the balloon through the pinhole. (The alligator clip swings down suddenly when the balloon is empty.)

When the valve handle is turned 180° counterclockwise from the position shown, gas from the syringe can enter the balloon.

you must get all the air out of the balloon. Next, the syringe has to be filled with a given volume of gas which is then transferred to the balloon. Finally, you measure the time for the gas in the balloon to effuse through the pinhole.

Deflating the Balloon.
To get all the air out of the balloon, turn the valve handle to the position shown in Fig. 19.12 so that the syringe is connected to the balloon and both are closed off from the air. As you pull the piston out, you will see the balloon flatten as nearly all the remaining air flows into the syringe. Turn the valve clockwise 180° to the position shown in Fig. 19.13 to prevent air from reentering the deflated balloon.

Filling the Syringe.
To fill the syringe with air, disconnect the valve from the syringe and pull the piston out, drawing in air, to exactly the 35 cm³ mark.

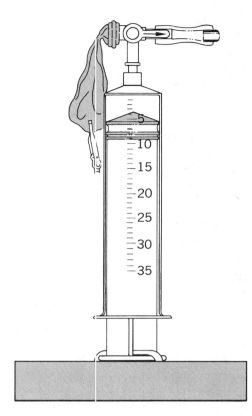

Fig. 19.12 The valve handle in position for completely deflating the balloon by pulling the piston away from the valve.

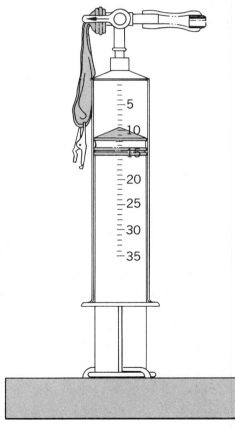

Fig. 19.13 The position of the valve before the syringe is filled with gas.

19.9 Experiment: The Effusion of Different Gases

Filling the Balloon.
To get the air in the syringe into the balloon, reconnect the valve to the syringe. Turn the valve handle counterclockwise 180° so as to reconnect the balloon with the syringe as shown in Fig. 19.14, and then push the piston all the way in.

Measuring the Effusion Time.
You are now ready to measure the effusion time for the air in the balloon. Begin the effusion by turning the valve handle 90° clockwise to connect the balloon to the pinhole (Fig. 19.15), and measure the time from the turning of the valve until the alligator clip falls.

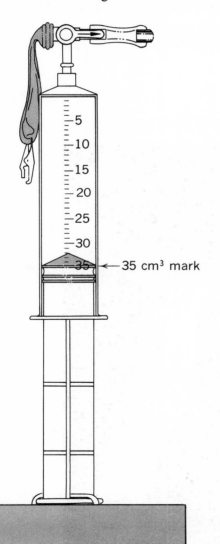

Fig. 19.14 The apparatus just before gas is transferred from the syringe to the balloon.

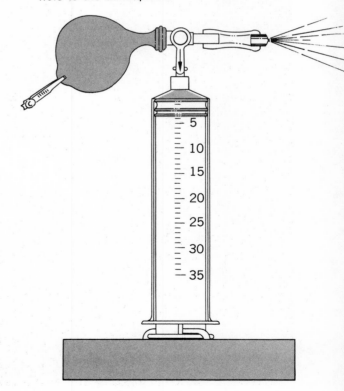

Fig. 19.15 To measure the effusion time, the valve is positioned as shown so that gas from the balloon can slowly leak through the pinhole to the atmosphere.

Make at least five runs and calculate the average effusion time.

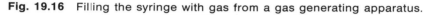

To measure the effusion time of hydrogen gas, first completely deflate the balloon as you did at the beginning of the experiment. Next, prepare a source of hydrogen as shown in Fig. 19.16 by adding about 15 cm^3 of dilute hydrochloric acid to about 4 g of zinc in a test tube. Let the generator run about 30 seconds to be sure all the air has been driven out of the test tube before proceeding with the next step.

To fill the syringe with hydrogen, disconnect the valve from the syringe and push the piston all the way in to expel all the air from the syringe. Then connect the syringe to the hydrogen generator as shown in Fig. 19.16. Allow the pressure of the gas from the generator to fill the syringe with hydrogen by pushing the piston to just beyond the 35 cm^3 mark.

Fig. 19.16 Filling the syringe with gas from a gas generating apparatus.

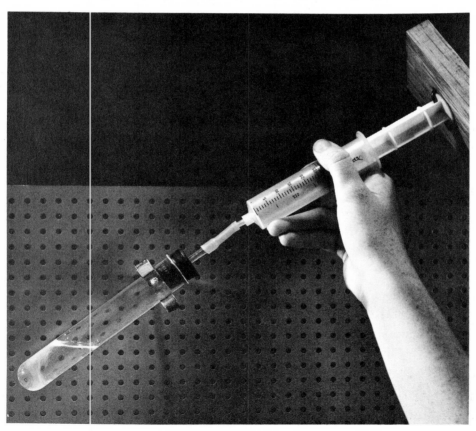

Disconnect the syringe, push the piston back to the 35 cm³ mark, and reconnect the valve to the syringe. If these three steps are done in rapid succession, there will be little chance for hydrogen to diffuse out of the syringe and for air to diffuse into the syringe.

Now, after turning the valve 180° clockwise to the position shown in Fig. 19.14, you can fill the balloon with the hydrogen from the syringe and proceed to measure the effusion time.

Make several runs. To reduce the time needed to fill the syringe with hydrogen, recharge the hydrogen generator with hydrochloric acid before each run. After five or more runs, calculate the average effusion time.

Next, measure the effusion time of carbon dioxide. This gas can be generated by placing about 10 cm³ of water in the generator test tube and adding about one-fourth of an Alka-Seltzer tablet, or by adding 10 g of marble chips to half a test tube of dilute hydrochloric acid. It is advisable to make a final series of runs using air again. Why?

How do the masses of molecules of hydrogen (H_2) and carbon dioxide (CO_2) compare with the mass of the molecules in air? You can assume that all the molecules in the air have the same mass as nitrogen molecules (N_2), since air is about 80 percent nitrogen, and the oxygen molecules which make up most of the remainder have masses close to those of nitrogen molecules. Do the effusion times you have measured bear out the prediction we have made about the qualitative relation between average molecular speed in a gas and its molecular mass?

16† The hole through which gases effuse in Expt. 19.9 has a diameter of about 10^{-4} m and the gas molecules have a diameter of about 10^{-10} m.
   a) What is the approximate ratio of molecular diameter to hole diameter?
   b) How large an opening should you use to represent the hole in Expt. 19.9 if you use marbles of 1 cm diameter to represent gas molecules?

For Home, Desk, and Lab

17 Suppose the aluminum cylinder in Expt. 19.1 had twice the mass of the one you used. How would this affect the temperature rise of the cylinder?

18 Suppose you replaced the aluminum cylinder in Expt. 19.1 by a copper cylinder of the same mass. How would this affect the temperature rise of the cylinder?

19 In Expt. 19.1, how would the temperature rise in the aluminum cylinder compare with the rise you actually observed if you stopped the wheel
   a) very slowly?
   b) very quickly?

20 In Expt. 19.1 we assumed that no heat was generated by the nylon string unwinding on the hub. What could you do to find out if this assumption is correct?

21 If the wheel of Expt. 19.1 is not stopped with the brake but is allowed to spin until it stops, where do you think the kinetic energy goes?

22 Describe the energy changes that occur while the mass shown in Fig. 17.9 falls from its starting position to its lowest position.

23 How will the kinetic energy of a mass M moving with a speed v be affected if
   a) M is doubled while v remains constant?
   b) v is halved while M remains constant?
   c) both M and v are doubled?
   d) M is doubled and v is halved?
   e) M is halved and v is doubled?

24 A student reported that in an experiment like Expt. 19.4 he let a mass of 0.50 kg fall 1.00 m. The rim speed of the wheel was 1.4 m/sec. Could the wheel have been like the one you used in Expt. 19.4?

25 Two carts move in a straight line, both at the same speed. Cart No. 1 has three times as much mass as Cart No. 2. What is the ratio of their kinetic energies?

26 The two carts in Problem 25 now move at different speeds, such that the ratio of the speed of Cart No. 1 to that of Cart No. 2 is 3.0. What is now the ratio of their kinetic energies?

27 If the cart in Fig. A is released, how high will it go on the incline at the right?

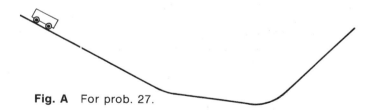

Fig. A For prob. 27.

28 Two identical cars move down the same hill with their brakes on at constant speed, but one moves faster than the other. In which car will the brakes get hotter?

29 Two stones, one with mass M, the other with mass $2M$, are dropped from the same height and fall to the ground. How do their kinetic energies compare just before they strike the ground?

30 Which object has the greater kinetic energy, a stone with a mass of 40 g falling with a speed of 10 m/sec or a 400-g stone falling with a speed of 1 m/sec?

31 A sky diver falls with increasing speed for a while, but he soon reaches a constant speed (about 100 miles per hour). How can you explain the fact that after he reaches a constant speed, he is constantly losing gravitational potential energy without gaining any more kinetic energy?

32 An object is suspended by a string from the ceiling. When it is pulled to one side and released, it swings back and forth.
   a) When is its potential energy increasing? Decreasing?
   b) When is its kinetic energy increasing? Decreasing?

33 To stop a car moving on a horizontal road, you have to apply the brakes. To keep a car at a constant speed while it is descending a hill, you must also apply the brakes.
   Consider two identical cars of mass 1,000 kg. One moving horizontally is brought to a stop from an initial speed of 25 m/sec (about 55 mi/hr); the other descends a hill 300 m high (about 1,000 feet) at constant speed and is not stopped. In which set of brakes will the heat released be larger?

34 From your data for Expt. 17.2, Gravitational Potential Energy as a Function of Mass, calculate the ratio of the change in kinetic energy of one of the falling masses to its change in gravitational potential energy. If you did not record the falling time, assume it to be 10 sec.

35 In determining the value of the constant in Sec. 19.5 we neglected the kinetic energy of the hub. The hub, including both the metal part and the black, hard-rubber part on which the string is wound, has a mass of 0.20 kg.
   a) Assuming that all this mass moves at the same speed as the outside of the hub on which the string is wound, calculate the kinetic energy of the hub, when the kilogram mass falls 1.00 m.
   b) Would its actual kinetic energy be more or less than the value you have calculated?
   c) Is the kinetic energy of the hub negligible?

36 In Expt. 19.1, Heat Generated by a Rotating Wheel, approximately 3.0 kg fell a distance of 1.0 m in putting a 5.0-kg wheel in motion.
   a) What was the gain in kinetic energy of the wheel and its final speed? (Use a table of square roots, or estimate the square root of the square of the speed.)
   b) Use the speed of the wheel to find the speed of the falling mass. (Refer to the data in Sec. 19.6.)
   c) What was the kinetic energy of the falling mass? What percent was this of the kinetic energy of the wheel?

37 Suppose we wrap the string in Expt. 19.1 around the rim of the wheel instead of around the rubber hub. Would your temperature rise for a given change in height be the same as before? If not, which would be the smaller? Why?

38 A wheel with most of its mass on the rim and a solid disc of the same radius and mass make one revolution in the same time. Which has the greater kinetic energy?

39 One 1,000-cm³ container of hydrogen has a temperature of 100°C. Another 1,000-cm³ container of the same gas of the same density is at a temperature of 0°C. In which container would you expect
   a) the average speed of the molecules to be greater?
   b) the total kinetic energy to be greater?

40 Two identical jars are filled with air at room temperature. One jar is placed in a freezer, the other in a pan of boiling water.
   a) What do you predict about the average speeds of the molecules in the two jars?
   b) Do you predict that all the molecules of gas in one jar are moving faster than all those in the other jar?

41 a) Use Table 12.1 to calculate the mass in grams of a molecule of ammonia (NH_3) and a molecule of hydrogen chloride gas (HCl).
   b) If ammonia and hydrogen chloride gas are released into opposite ends of a long, hollow glass tube at the same time, after a while they combine to form a white cloud of ammonium chloride. Would you expect the white cloud to form first in the middle, nearer the ammonia end, or nearer the hydrogen chloride end of the tube? Try it.

20 The Conservation of Energy

20.1 Laws and Definitions

In your studies in Physical Science you have encountered two kinds of mathematical relations: laws and definitions. Although both are often expressed by equations, they are of very different character. As was stated in Sec. 2.13, laws of nature are guessed generalizations based on experiments. Consider the law of conservation of mass. On the basis of measurements of masses before and after a reaction took place, the generalization was made that:

$$\text{Mass (before)} = \text{mass (after)}$$

Another example is Boyle's law: if the volume of a gas is decreased by a certain factor, the pressure is increased by the same factor (Sec. 10.4). This relation is expressed by the equation

$$\text{Volume} = \frac{\text{constant}}{\text{pressure}}$$

We predicted Boyle's law from the atomic model and compared the predictions with experiments in which changes in volume and pressure of a gas were actually measured. We found that Boyle's law indeed summarizes the experimental data very well for pressures that are not too high. At high pressures there are deviations from the law (Sec. 10.8).

Contrast the equations expressing the law of conservation of mass and Boyle's law with equations relating density, mass, and volume:

$$\text{Density} = \frac{\text{mass}}{\text{volume}}$$

The relation expressed by this equation is not based on any experiments, nor can it be disproved by any experiment. It simply gives a name to the quantity, mass per unit volume; it defines the term "density." Similarly, think of the relation between heat capacity, quantity of heat, and temperature change:

$$\text{Heat capacity} = \frac{\text{quantity of heat}}{\text{temp. change}}$$

This equation also is not a generalization based on experiments. It only defines what we mean by the term "heat capacity." As you will see in the next section, the distinction between laws and definitions is significant in our study of energy.

1   Water boils at 100°C and freezes at 0°C. The melting point and freezing point of water occur at the same temperature. Which statement is based on experimental evidence?

2† Is the relation electrical work = current × time × voltage a definition or a law of nature?

A Review of Energy Changes   20.2

Suppose someone had asked you after your study of dry cells in Chapter 13 if there is any connection between a rising mass and the mass of zinc dissolved in a dry cell. You would probably have answered "no." Now you would answer "yes," since in the meantime you have done several experiments which showed that these processes have something in common. This is because changes in energy occur both when zinc dissolves in a battery and when a mass rises. In fact, you have done additional experiments which involved processes that apparently were unrelated, while in fact changes in energy occurred in each of them. Let us review our reasoning and conclusions about the energy changes that took place in these experiments.

The starting point was with the conclusions drawn from the motor experiments in Chapter 16. A given amount of work done on an electric motor produces an equal amount of heat whether the motor's shaft is held fixed or is free-running (Fig. 20.1). When the motor lifts a mass, the heat produced is less than the work done. Some heat is "missing." Part of the electrical work can either lift a mass or heat up the motor. The missing heat is recovered, however, if the mass is then permitted to fall back to its original position, while spinning the shaft of the motor (Fig. 20.2). In other words, the falling mass produces the same amount of heat that could

20.2 A Review of Energy Changes

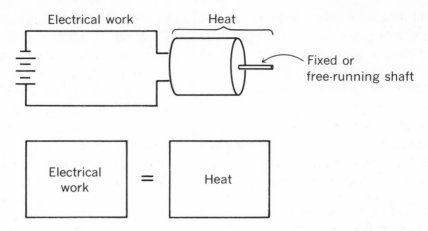

Fig. 20.1 In a motor whose shaft is fixed or free-running, the heat produced in the motor is equal to the electrical work done on it.

Fig. 20.2 When a motor lifts a weight, the heat produced is less than the electrical work. When the weight falls back to its starting position, the missing heat is recovered.

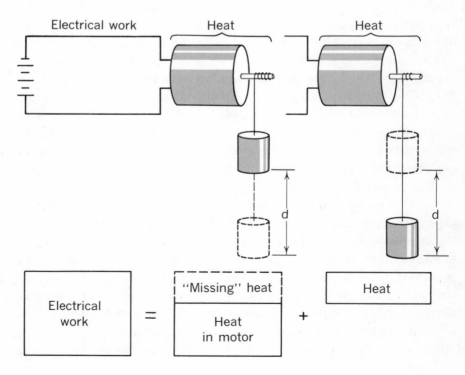

be produced in a heater by the electrical work required to lift the mass in the first place.

A given amount of work can either lift a mass or heat up an object, and we can trade one change for the other by letting a falling mass produce heat. Therefore, we say that both are equal changes in something called *energy*. However, since lifting a mass and producing heat are distinctly different processes, we give the two energies different names: gravitational potential energy and thermal energy.

We then decided to say, by definition, that the changes in gravitational potential energy and thermal energy are equal when they are produced, or could be produced, by the same amount of electrical work. This equality cannot be checked by experiment, since it is part of our definition of gravitational potential energy.

We introduced the idea of kinetic energy in a similar way: A slowly falling mass can generate a given quantity of heat. The same mass falling the same distance can also set a wheel in motion without generating any significant amount of heat. However, when the wheel is stopped, the quantity of heat produced is equal to that which would be produced by the falling mass directly. We therefore identify the change in the motion of the wheel as a change in energy, calling this form of energy kinetic energy. We defined the gain in kinetic energy of the wheel as equal to the loss of potential energy of the falling mass, and this equality also is not subject to experimental verification.

The question which now arises is: Once we define the changes in one form of energy in terms of changes in another form of energy, will *any* chain of successive changes produce the same net effect, independent of the intermediate steps?

Consider the following example. A bicycle wheel is connected to an electric motor, and the leads from the motor are connected to a hydrogen cell (Fig. 20.3). At the beginning the wheel is spinning and turning the shaft of the motor, which generates a voltage. You may recall that this was also the case in Expt. 16.5 when the falling mass turned the motor shaft. When charge flows through the hydrogen cell, water is decomposed, and a mixture of hydrogen and oxygen gas collects in the test tube, and the water in the cell warms up. Charge also flows through the motor, heating it. While these changes are taking place, the bicycle wheel slows down and finally comes to a stop. When it stops, we ignite the hydrogen-oxygen mixture and produce thermal energy. In the overall process the kinetic energy of the wheel decreased, and the thermal energy increased.

We could, of course, stop the wheel by letting it rub against an aluminum cylinder as you did in Expt. 19.1. Would the increase in the

20.2 A Review of Energy Changes

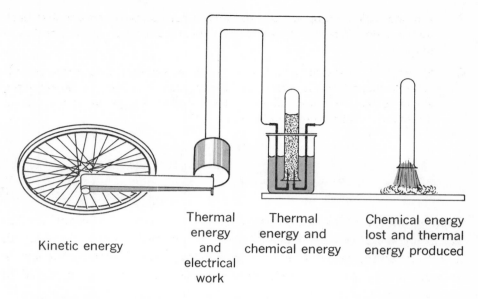

Kinetic energy | Thermal energy and electrical work | Thermal energy and chemical energy | Chemical energy lost and thermal energy produced

Fig. 20.3 Kinetic energy from the wheel is converted into thermal energy and electrical work in the motor. The electrical work produces heat and chemical energy (atomic potential energy) in the hydrogen cell. When the mixture of gases in the hydrogen cell is ignited, chemical energy is converted into heat.

Fig. 20.4 Is the increase in thermal energy of the motor, hydrogen cell, and reacting hydrogen and oxygen equal to the decrease in kinetic energy of the wheel?

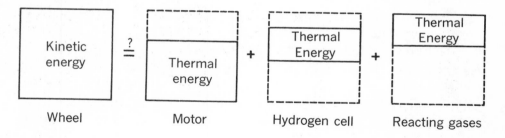

Wheel | Motor | Hydrogen cell | Reacting gases

thermal energy of the cylinder equal the increase in thermal energy of the motor, the hydrogen cell, and water produced by the reacting gases? Or, in other words, in the chain of energy transfers shown in Fig. 20.4, would the *increase* in thermal energy equal the *decrease* in the kinetic energy of the wheel? If we tried to answer these questions experimentally, we would encounter quite a few technical difficulties, and so we shall not attempt it. Another chain for which it is easier to carry out all the measurements is the topic for the next experiment.

3   A rubber ball is dropped 1.0 meter onto a hard surface and bounces back up 0.9 meter. Describe the energy changes as a function of the position of the ball.

4   Steam under pressure is put into a cylinder closed by a piston. The cylinder is well insulated, so that little thermal energy is lost to the outside. When the steam pushes the piston, the steam loses thermal energy. In terms of steam molecules, what happens during the push?

Experiment
A Series of Energy Changes 20.3

Suppose that a 1.00-kg mass is attached to the end of a spring so that the spring is taut but not stretched at all. If the mass is suddenly released, it will fall, losing gravitational potential energy and stretching the spring to some length d (Fig. 20.5a). Now suppose the same spring is tied to

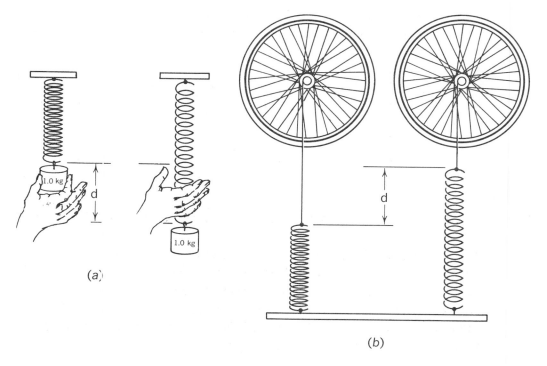

Fig. 20.5 (a) A 1.00-kg mass is attached to a spring and held so the spring is not stretched. It is suddenly released and allowed to fall, stretching the spring (as was shown in Fig. 17.9). In falling, it stretches the spring an amount d. (b) The same spring is then connected to the axle of a bicycle wheel, and the wheel is turned by hand until the spring is stretched the same amount d as it was in (a).

a length of string that is wound around the drum of the wheel used in Expt. 19.1. The wheel is then rotated by hand so that it winds up the string and stretches the spring by an amount equal to the maximum stretch it had when the 1.00-kg mass was hung on it (Fig. 20.5b). Then, if the wheel is released, the spring will contract, speeding up the wheel. What energy changes do you predict will take place from the beginning to the end of the experiment?

Measure the maximum stretch of the spring when a 1.00-kg mass is attached to it and is allowed to fall suddenly. What do you predict will be the final speed of the wheel? What assumptions have you made?

Finish the experiment to check your prediction. Is the final speed of the wheel the same as you would have expected it to be if you had let the falling mass speed up the wheel directly instead of indirectly by means of the spring?

Experiment
20.4 The Energy Associated with Light

A flashlight bulb like the one you used earlier in this course consists of a fine metal wire enclosed in an evacuated glass bulb. When electric charge flows through the wire, it heats up. Unlike the heaters you have used in your experiments, a light bulb gives off light as well as heat. Does the electrical work done on the bulb equal the heat produced in it? You can answer this question experimentally with the apparatus shown in Fig. 20.6. A light bulb is submerged in water in a styrofoam container with plastic windows which let most of the light from the bulb escape. Heat from the bulb flows into the water and increases its thermal energy. The electrical work and the change in thermal energy in the container are measured and then compared. The experiment can then be repeated with aluminum foil covering the light bulb so that very little light escapes.

You can measure the electrical work in the standard way. To determine the increase in thermal energy, be sure you know accurately the mass of water in the container. (The total heat capacity of the light bulb and container is very small, and if about 75 g of water is used, it can be neglected.)

Connect the apparatus to the battery shown in Fig. 20.7 and record the current and voltage every 30 sec. Start with slightly precooled water, say 1°C below room temperature, and keep the bulb lighted until the water is 1°C above room temperature. Why should you stop at 1°C? When you make a temperature reading, be sure to mix the water by alternately

Experiment: The Energy Associated with Light 20.4

Fig. 20.6 Electrical work is supplied to a light bulb submerged in water in the plastic container. The electrical work and the thermal energy produced in the container can be measured in the usual way and then compared. The experiment can be repeated with aluminum foil (at left, above) covering the light bulb so that very little light can escape. The heat capacity of the foil is negligible.

Fig. 20.7 The battery used in Expt. 20.4.

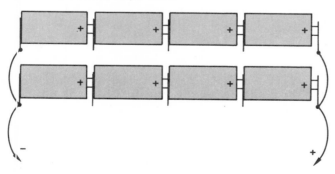

swirling the water and reading the thermometer until the temperature reading is steady. What is the ratio of the increase in thermal energy to the electrical work?

Repeat the experiment with the bulb covered with aluminum foil. How does the ratio of the increase in thermal energy to the electrical work when the bulb was covered compare with the ratio when the bulb was uncovered?

How is this experiment related to the experiments using an electric motor to lift an object? (Expt. 16.4 and 16.5.)

20.5 Radiant Energy

Applying the reasoning which we developed earlier, we can now associate light with another form of energy, namely *radiant energy*. The amount of radiant energy contained in a given beam of light is then defined to be equal to the increase in thermal energy of an object which absorbs the beam of light.

The existence of radiant energy suggests sequences of energy changes in addition to those we have discussed. Here is an example: A strip of magnesium is burned in a darkened room where there is an uncovered photographic film. When it burns, it combines with oxygen and nitrogen in the air, forming magnesium oxide and magnesium nitride. It burns with a bright, white light. The radiant energy from the burning magnesium is absorbed by the walls, ceiling, and floor of the room and by any objects in the room, including the photographic film. The absorbed radiant energy is mostly changed into thermal energy. However, some of the energy absorbed by the film does not become thermal energy. Instead, it "exposes" the film. In this process radiant energy is changed to chemical energy in the decomposition of silver bromide in the film to silver and bromine. In the overall process the chemical energy of the magnesium and the air it combines with is changed to thermal and radiant energy. Most of the radiant energy is converted to thermal energy, but some of it is converted back into chemical energy by the photographic film.

Here is another example, one that involves a device called a radiometer (Fig. 20.8). When a flashlight beam is directed at it, the four vanes of the radiometer begin to revolve. In this case chemical energy in the flashlight battery is exchanged for thermal energy and radiant energy in the flashlight bulb. Some of this radiant energy is absorbed by the radiometer, producing both thermal energy and kinetic energy of the moving vanes.

Fig. 20.8 A radiometer. Radiant energy falling on the four vanes is strongly absorbed by the black sides of the vanes but is mostly reflected by the shiny sides. The unequal heating produced on the two sides of each vane causes the vanes to revolve.

The Law of Conservation of Energy 20.6

On the basis of many measurements involving all possible kinds of energy changes, we can state that in all reactions known today energy never disappears and energy is never created from nothing. When a reaction takes place, one form of energy can increase, but this is always at the expense of an equal decrease in another form (or forms) of energy. This generalization is known as the law of conservation of energy.

It is interesting to note that the idea that the total energy of a system does not change during a process was already expressed over 100 years ago—before the word *energy* itself was used. It was known that a slowly falling object could *produce* motion, and that a rapidly moving object could, through friction, *produce* heat. Heat in its turn could lift an object or produce motion (through the medium of a steam engine). However, it was not understood that a single quantity (what we call "energy") was being converted, unaltered in amount, from one form to another in these processes. Then, in a short period of about ten years around 1845, rapid strides were made in the understanding of these processes. In 1847, largely as a result of careful measurements by an English brewery owner, James Joule, the law of conservation of energy was arrived at in its general form by both Joule and von Helmholtz, a German physician.

Here is a quotation from a lecture given by Joule in 1847. Some of the terms he used are no longer used today. To make the reading easier, we have added the modern terms in brackets.

20.6 The Law of Conservation of Energy

.... All three, therefore—namely, heat [thermal energy] living force [kinetic energy] and attraction through space [potential energy] (to which I might also add LIGHT, were it consistent with the scope of the present lecture)—are mutually convertible into one another. In these conversions nothing is ever lost. The same quantity of heat [thermal energy] will always be converted into the same quantity of living force [kinetic energy]. We can therefore express the equivalency in definite language applicable at all times and under all circumstances. Thus the attraction of 817 lb. through the space of one foot [change in potential energy] is equivalent to, and convertible into, the living force [kinetic energy] possessed by a body of the same weight of 817 lb. when moving with the velocity of eight feet per second, and this living force [kinetic energy] is again convertible into the quantity of heat [thermal energy] which can increase the temperature of one pound of water by one degree Fahrenheit.

There is another form of energy of which Joule was unaware: *nuclear energy,* sometimes inaccurately called "atomic" energy. You may recall that radioactive polonium was used in determining the mass of a helium atom in the experiment in Chapter 9 of *Introductory Physical Science,* and that it produced a blue glow in the sealed quartz tube in which it decayed. The polonium also produced so much thermal energy that the quartz tube stayed warm—almost hot—during the many weeks in which the polonium decayed. Both the radiant energy and the thermal energy came from the centers or nuclei of the polonium atoms. Nuclear energy, like chemical energy, kinetic energy, etc., can be converted into other forms and must, of course, be taken into account in the overall conservation of energy.

We shall not study nuclear energy in any detail in this course. It is mentioned only to complete the picture of the conservation of energy.

5   Suppose in the quotation from Joule's lecture you replace 817 lb by 400 kg, 1 foot by 0.3 meter, and 1 pound of water by 0.5 kg of water. What would be the velocity of the 400-kg mass in m/sec and the rise in temperature of the 0.5 kg of water in °C?

6   A satellite S orbiting the earth follows the path shown in Fig. A. Will the speed of the satellite be greater at point 1 or at point 3? How does the speed at point 2 compare with the speed at point 4? On what do you base your answers?

Fig. A   For prob. 6.

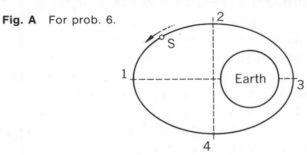

Nonreversible Processes: The One-Way Street 20.7

From your study of energy you would be very skeptical if someone claimed that he saw a process take place in which there was a net gain in energy. You would not believe it, for example, if you were told that a car started moving on a level street without its engine running and without being pushed. Neither is it possible that there is a saw that cuts wood indefinitely without being connected to a supply of energy. Imaginary devices designed to do things of this kind are called perpetual-motion machines. They never work. There are no machines or processes that violate the law of conservation of energy.

But does this mean that any process is possible as long as it does not violate the law of conservation of energy?

There are many processes in which energy would be conserved but which do not happen. Suppose you have a closed box divided in half by a partition, with air in one half and a vacuum in the other half. You know what will happen if the partition is removed: The air will spread evenly throughout the box. But would the reverse process ever occur? That is, would the gas ever, by itself, all crowd back into the same half of the box, even though this does not contradict the law of conservation of energy?

Experiment
A Disc Gas with Few "Molecules" 20.8

You can use the "disc gas" machine described in Sec. 19.8 to represent the gas in the partitioned box described in the last section. A wooden barrier will serve to keep the discs initially in one half of the table. Once you remove the barrier, the discs will move over the entire table. Occasionally it may happen that for a short while all the discs will again be in the same half of the table in which they were originally. Does the average time that will elapse until this happens depend on the number of discs?

To find out, start with one disc and record the total time it takes for this disc to return 10 times to the initial half of the surface. (Be sure the disc is moving at a reasonable speed when you start timing.) From this value calculate the average time it takes for this disc to return to the initial half of the table. Now find the average time it takes for two discs to again be in the same half of the table in which they were originally. Repeat this procedure using three discs, four discs, etc.

Draw a graph of the average time it takes for all the discs to return to the half from which they started, as a function of the number of discs. What do you conclude about the average time it takes for all the discs to return to the half of the machine they started from as the number of discs is increased?

_____   _____   _____

In this experiment you were dealing with less than 10 disc "molecules." Suppose the table were large enough to accommodate 100 discs. It is extremely unlikely that you would ever see them again in one half of the table, even if you spent your entire lifetime watching.

Now consider a cubic centimeter of air. It contains 10^{20} gas molecules. It is so highly unlikely that this many molecules in the partitioned box described in Sec. 20.7 would ever return to the half from which they originated that you would never expect this to occur.

20.9 Nonreversible Processes and Large Numbers of Atoms

The fact that any sample of matter as small as the head of a pin contains of the order of 10^{20} atoms is also at the root of other nonreversible processes.

Suppose you drop a ball of plasticine onto a table. It sticks to the table and does not bounce. As it hits, its kinetic energy is transformed into thermal energy. Energy is conserved. Now imagine the reverse process, in which the ball loses some thermal energy which is converted into kinetic energy of the ball. That is, the ball cools off a little by itself without heating up the surroundings and starts moving upward. Even though a temperature drop in the ball of only $0.02°C$ would produce enough kinetic energy to lift the ball almost 2 meters, such a process never occurs. If you wanted to see what it would look like, you would have to take motion pictures of a falling plasticine ball and run the film backward through the projector.

Let us look at the whole process on the atomic scale. When the ball is at rest in your hand, its atoms vibrate equally in all directions. During the fall the downward motion of the atoms is slightly faster and the upward motion is slower. (Otherwise, the ball as a whole would not be falling.) After the impact the ball is at rest, but at a higher temperature. This means that its atoms are again vibrating equally in all directions but more vigorously than before. For the plasticine ball to lift itself off the table, all the atoms must increase their velocity in the upward direction at about the same instant. In principle this can happen. But it is just as improbable as 10^{20} molecules of gas in a partitioned box ever returning to one half of the box.

Now contrast a plasticine ball with a rubber ball. When the latter hits a hard table, most of the kinetic energy of the ball is transformed into elastic potential energy, rather than thermal energy. This does not increase the random vibrations of the atoms. It decreases their separation only in the vertical direction. The atoms then push apart again and the ball moves up. However, even a rubber ball does not bounce back to the height it was dropped from. Some kinetic energy is transformed into thermal energy, indicating that atomic vibrations were affected.

The bouncing of the almost ideal rubber ball illustrates a general principle: Whenever there is a change of kinetic energy to potential energy or *vice versa*, there is some increase in thermal energy at the expense of other forms of energy. Thus, no cart can roll down one inclined plane and up another one to the original height, no matter how good the bearings are; no motor can lift a weight without warming up.

Sometimes these effects are very small and therefore of little practical concern. However, they may not be negligible over a long time. The presence of the moon and the rotation of the earth are responsible for the occurrence of the tides—that is, the back-and-forth motion of huge quantities of water. The major energy changes during this process are the exchange of potential energy and kinetic energy. But like other such processes they are not completely reversible. The water of the oceans warms up ever so slightly. Since energy is conserved, the increase in thermal energy must come at the expense of a decrease of another form of energy. A small decrease in the rate of rotation of the earth around its axis could provide this thermal energy. Indeed, there is evidence that the length of a day increases about 0.002 sec each century. This is very little, but over a long time it will be significant.

For Home, Desk, and Lab

7   We can measure voltage across a copper wire kept at a constant temperature, and the current through it. Which of the following two relations is a law and which is a definition?

     a) $\dfrac{\text{Voltage}}{\text{Current}} = \text{resistance}$

     b)   Resistance = constant

8 An object is hung on a string which is fastened securely and wound around the hub of a bicycle wheel like the one you used in Chapter 19. When the object is released, it falls and turns the wheel, but the string is too short to allow the object to reach the floor. Describe the energy changes in the object and the wheel until all motion ceases.

9 In an air gun, air is first compressed; when the gun is fired, the compressed air speeds up the bullet. The bullet travels through the atmosphere and finally embeds itself in a wooden target. What changes of energy occur? Where do they occur?

10 The aluminum foil you used in Expt. 20.4, The Energy Associated with Light, had a mass of about 0.2 g. How does its heat capacity compare with that of the water?

11 If the film in a motion-picture projector accidentally stops when the projection light is on, the film very often starts burning. Would a daylight scene of a desert catch fire more quickly than a scene of a rainy evening?

12 A filter paper exposed to the sunlight will not catch fire. However, if a magnifying lens of the same diameter as the filter paper is held above the paper, it does catch fire. Why?

13 A disc-gas machine is set up with the wooden barrier keeping the discs initially in three-fourths of the table instead of one-half. How will the average time for the discs to return to the original three-fourths of the table compare with the time in experiments where the discs returned to half of the table? Try it with the largest number of discs you used in Expt. 20.8.

14 From how far away can you blow out a lighted candle? Explain in terms of the motion of air molecules why you cannot blow out the candle from across the room.

15 When you bend a piece of copper wire back and forth for a period of time, the wire will get hot. Describe this process in terms of the motion of atoms.

16 A student arranges an electric motor to drive an electric generator. He proposes to have the output from the generator drive the motor, so that once started, the system will continue to run. Do you expect his attempt to be successful?

Energy: A Global View   21

Throughout this course we have studied energy changes in the laboratory. We devised experiments to measure these changes, and from our measurements we were lead to the conviction that energy is always conserved. Energy transformations, of course, are not confined to the laboratory. On a much larger scale, we see them every day in the world around us. Some of these energy changes take place by themselves—for example, those that result in the kinetic energy of the winds and ocean currents, the energy locked up in fossil fuels, and the energy that comes from the food that living things consume. Others, like those that result in the potential energy stored behind dams and the kinetic energy of airplanes, are due to man's ingenuity in controlling the energy occurring in nature. In this chapter we shall trace some examples of both natural energy changes and energy changes controlled by man.

Absorption of Radiant Energy from the Sun   21.1

Very nearly all the energy changes that take place on the earth involve energy that arrived as radiant energy from the sun.

Not all the radiant energy from the sun that arrives in the upper atmosphere is absorbed by the atmosphere and the earth. Much of it is reflected back into space without being absorbed. Most of the radiant energy from the sun that is absorbed by the land and the oceans changes to thermal energy. Both the earth and the atmosphere also radiate energy back into space. The amount of radiation absorbed by the earth and the atmosphere is very nearly equal to the amount that is re-radiated to space, because we observe no significant change in the average temperature of the atmosphere over a period of many years. If the earth absorbed more radiation than it radiates back into space, its average temperature would steadily rise.

21.2 Energy Changes in a Hurricane

A major effect of the absorption of the sun's radiation is the evaporation of water from the surface of the oceans. In terms of energy changes this means a decrease in radiant energy and an increase in the atomic potential energy of the water molecules (heat of vaporization of water). Eventually all the water vapor condenses, and much of the resulting thermal energy is transformed into the kinetic energy of the winds, waves, and currents associated with storms. These forms of kinetic energy finally turn into thermal energy as the storms die out.

To see how the heat of vaporization of ocean water is converted into the kinetic energy of winds and waves, we shall follow the major energy transformations that occur in a large and violent storm of the type called a hurricane. How hurricanes get started is not completely understood. For our purposes it is only necessary to describe a fully developed hurricane.

A hurricane is a vast swirl or eddy of air having a diameter of approximately 300 km (200 miles). In a hurricane the air spirals inward (Fig. 21.1) around the center of the storm with surface winds of "hurricane force," winds of speeds greater than 120 km/hr (74 mi/hr).

The air also moves upward as it spirals around the "eye" of the hurricane. This is shown in Fig. 21.2. As the ocean air, saturated with water vapor, flows up through the storm, it expands because of the drop in atmospheric pressure with increase in altitude. It therefore cools as it rises, just the reverse of what happens when a gas is compressed (Sec. 10.5, Chapter 10).

When the rising air is cold enough, the water vapor starts condensing. The thermal energy produced by the heat of condensation slows down the cooling of the rising air. Therefore, at any given level above the altitude at which the condensation begins, the rising air is warmer than air surrounding the storm at that same level. If it is warmer, it is also less dense than the surrounding air, and so it continues to rise. As a result, more moist air flows into the storm horizontally near the ocean surface to take the place of the rising air (Fig. 21.2). As this moist air rises, it provides more heat of vaporization to keep the air flowing and the storm alive.

The moist air moving into the storm over the ocean's surface at high speed is the hurricane wind that does so much damage. This wind transfers, by friction, kinetic energy to the ocean surface, resulting in the large destructive waves and currents typical of a hurricane.

The energy of an average-size hurricane during its lifetime (about 9 days) is 10^{19} joules, ten times the energy produced in the world by man-made machines in the same time interval.

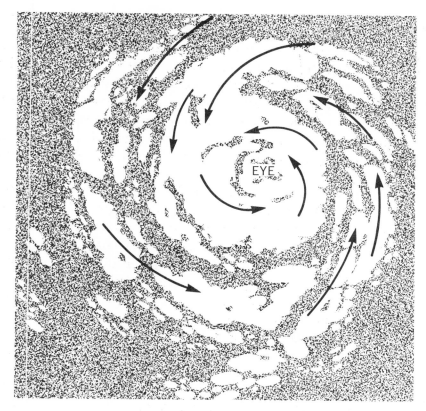

Fig. 21.1 Looking down on a hurricane from above. The winds and clouds spiral inward toward the cloudless, calm center called the "eye."

Fig. 21.2 A vertical "slice" through the center of a fully developed hurricane, showing how the air moves upward through the storm, out at the top, and downward in the eye. (In this diagram the height is exaggerated.)

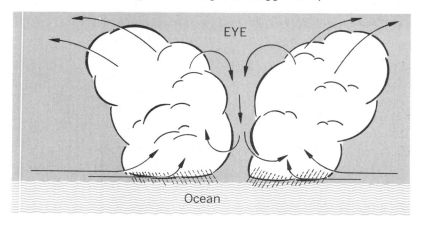

21.3 Photosynthesis

1. Figure A illustrates the water cycle, the process by which water moves from the earth's surface to the atmosphere and back again. What energy changes are involved in this cycle?

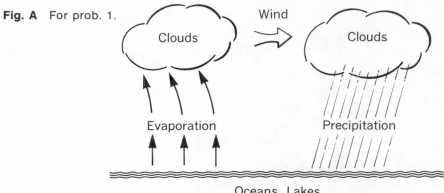

Fig. A For prob. 1.

2. The rain falling from a hurricane is very intense near the center but drops off to brief light showers near the outer limits of the storm. Suppose that a hurricane has a diameter of 400 km, and that the average rainfall is 10 cm per day over the entire area of the storm for its 10 day lifetime. What is the total heat of condensation (in joules) from the storm?

3. Why are there often only a few scattered clouds or no clouds in the eye of a hurricane? (See Fig. 21.2.)

4. Where in Fig. 21.2 does water vapor start to condense into cloud droplets?

5. When a hurricane moves inland from the sea, it rapidly dies out. Why do you think this is so?

21.3 Photosynthesis

Living things use only a very small fraction (about 0.2 percent) of the solar energy that strikes the earth's surface. This is the energy utilized by green plants in a process called photosynthesis, in which they convert the two simple compounds carbon dioxide and water into oxygen and carbohydrates (compounds made up of carbon, hydrogen, and oxygen atoms). This is illustrated in Fig. 21.3. Carbohydrates are compounds such as starches and sugars that are food for animals and man. The carbohydrates and oxygen produced by a plant have more chemical energy than the carbon dioxide and water used to produce them. This additional energy comes from some of the sun's radiant energy that is absorbed by the plant.

Since photosynthesis occurs only in green plants, all other forms of life ultimately depend on green plants for food; chemical energy stored in food is the energy source for all living things.

The carbon dioxide and water used in photosynthesis are replenished by the respiration carried on by plants and animals. During respiration, oxygen from the atmosphere is taken in by living organisms and reacts with compounds formed from digested food. The end products of long chains of complex reactions are always carbon dioxide and water. When carbon dioxide and water are expelled by an organism, they again become available for photosynthesis (see Fig. 21.4).

In many of the chains of reactions in which food is broken down into carbon dioxide and water, chemical energy, stored in the food and

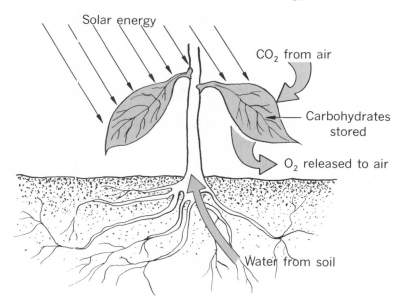

Fig. 21.3 In photosynthesis, radiant energy from the sun is absorbed by green plants, which convert water from the soil and carbon dioxide from the air into carbohydrates and oxygen.

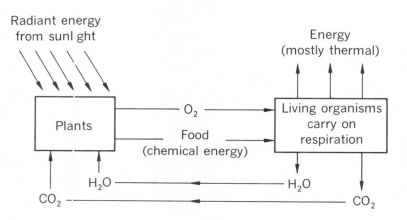

Fig. 21.4 The water-carbon dioxide cycle in photosynthesis and respiration. Of course, the same molecules of water and carbon dioxide that result from respiration are not necessarily used in photosynthesis again.

the oxygen, is transformed both to forms that can be utilized in life processes and to thermal energy. Eventually, however, all the chemical energy from photosynthesis goes down the "one-way street," becoming thermal energy that is no longer usable for photosynthesis.

21.4 Efficiency

Up to this point we have discussed only natural energy transformations. Now we shall consider some man-made machines designed for the purpose of converting energy from one form to another. The energy we get from the machine in the form we desire is called the output energy. We would like, of course, for the output energy to be equal in amount to the input energy, the energy we feed in.

However, when a machine is used to transform energy, the useful energy output from it is always less than the energy fed into it. This is confirmed by the results you obtained in the laboratory when you measured the heat produced in a motor that was used to lift a heavy bucket. Suppose the purpose of the experiment had been to lift the heavy bucket above the table rather than to find out if the heat equaled the electric work. The "missing heat" was the "useful" energy output; it equaled the increase in the gravitational energy of the bucket. The input was the electrical work, and it equaled the sum of the useful energy output plus the increase in thermal energy of the motor.

Because it is not possible to convert energy to a form that is useful without wasting a significant amount, it is important to know how the useful energy output compares with the total energy input—that is, how efficient the transformation is. For this purpose we define the efficiency of a machine as the ratio of useful energy output to energy input. We generally try to design machines with as high an efficiency as possible.

6 Sometimes, as in heating a house, we wish the transfer of energy to give nothing but thermal energy. In a furnace used to heat a building, chemical energy is transformed into thermal energy. Is all of this thermal energy utilized for the purpose of heating the building?

21.5 The Efficiency of an Automobile

About 20 percent of the total energy used each year in the United States is needed to run our more than 100 million automobiles. What is the efficiency of an automobile? One way of measuring it is based on some of the ideas you have used in your experiments. In Fig. 21.5(a) the jacked-up rear wheel of an automobile is connected to a large electric generator by means of a pulley belt. When the automobile is running, charge from the generator flows through a large electric heater (a resistor). Measurements of the voltage, the current, and the time—that is, the electrical work—show the energy transferred from the generator to the heater. If we know the efficiency of the generator, a comparison can be made between the energy released by the burning gasoline and the energy output from the wheels. (This energy output is used primarily to keep an automobile moving at the desired speed despite friction.)

For example, the heat of reaction of the burning of gasoline is 4.8×10^7 joules/kg, and 1 gallon of gasoline has a mass of about 2.5 kg. So, by burning 1 gallon of gasoline

$$4.8 \times 10^7 \text{ joules/kg} \times 2.5 \text{ kg} = 1.2 \times 10^8 \text{ joules}$$

of energy is released. Let us suppose that the automobile runs with the accelerator depressed enough so that one gallon of gasoline is consumed in 20 minutes. If the voltmeter and ammeter have average readings, say, of 320 volts and 48 amperes, then the electrical work is

$$(320 \text{ volts})(48 \text{ amp})(1{,}200 \text{ sec}) = 1.85 \times 10^7 \text{ joules}.$$

Fig. 21.5 (a) A way of measuring the efficiency of an automobile. The energy transfers are shown in (b).

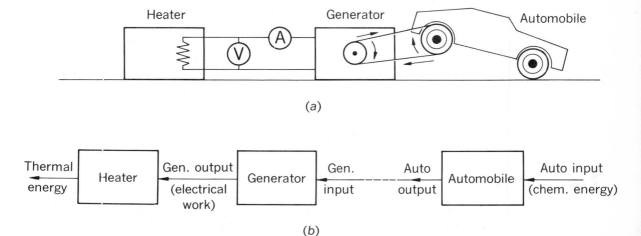

Let us assume that the generator is 80 percent efficient. Its input is the useful energy output of the automobile (Fig. 21.5(b)). It is

$$\frac{1.85 \times 10^7 \text{ joule}}{0.80} = 2.31 \times 10^7 \text{ joules}$$

The efficiency of the automobile would be the ratio of its useful energy output to its energy input, or

$$\frac{2.31 \times 10^7 \text{ joule}}{1.2 \times 10^8 \text{ joule}} = 0.19 = 19\%$$

This is the approximate efficiency of most automobiles. The nonuseful energy is in the form of thermal energy and is carried away by the cooling system.

There are other ways of measuring the performance of automobiles. From the point of view of economy you would be concerned with the number of miles per gallon of gasoline you can get from a car (which is related to its efficiency). You may be concerned, as well, with the reliability of an automobile in getting you from one place to another with minimum repairs. Or you may be concerned about the rate at which large numbers of cars deplete natural resources. The supply of petroleum is not unlimited, and therefore some other fuel may be necessary to supplement gasoline. The amount of pollution (also related to efficiency) that automobiles add to the environment is also important in judging their performance.

21.6 A Large-scale, Man-made Energy Converter

In a coal-burning electric power plant, chemical energy in the coal and in the oxygen of the air is transformed into thermal energy in the steam. Then, through the turbine and generator, it goes into the electrical work output. By measuring the thermal losses through the smokestack and the condenser (where the exhaust steam from the turbine is condensed into water, and the thermal energy from the heat of condensation is carried away), we can account for nearly all the energy fed into the system (Fig. 21.6).

At an electric power plant in Salem, Massachusetts, the investigation described above was carried out with the results shown in Table 21.1.

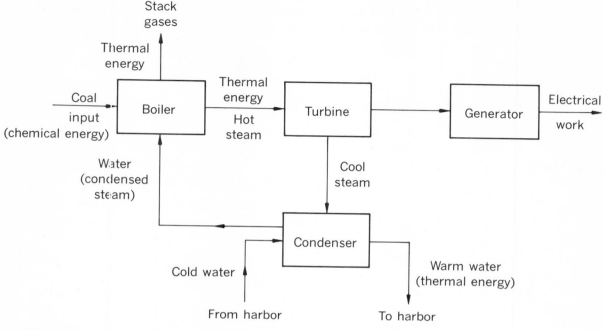

Fig. 21.6 A diagram of the energy flow in a coal-burning power plant.

Table 21.1 Billions of joules/hr

Input	Output	
From coal consumption, 770	Out the smokestack to atmosphere,	73
	From the condenser to harbor water,	381
	To electrical work,	308
770		762

We see that all the energy fed into the system is accounted for to within about 1 percent and that the efficiency of this particular power plant is

$$\frac{\text{Useful energy output}}{\text{Energy input}} = \frac{308 \text{ joule}}{770 \text{ joule}} = 0.40 = 40\%$$

This power plant is about twice as efficient as an automobile, but the thermal energy loss still constitutes a large percentage of the energy input. In many machines, whether for going to the moon or for erecting a skyscraper, more of the available energy goes directly into nonuseful thermal energy than into useful energy. However, there are a few devices that are over 90 percent efficient, such as the large water turbines used in hydroelectric plants and room-size electric generators, transformers, and motors.

7 Suppose the thermal energy released into Salem Harbor were used to heat buildings in Salem. How would this affect the efficiency of this power plant?

21.7 How Long Will the Major Energy Sources Last?

This year the world coal production will be more than 10 times what it was in 1870. The production of petroleum, which was negligible in 1870, is now more than 10^9 barrels a year and has been doubling every 10 years.

Although the same geological processes that formed the earth's deposit of fossil fuels (petroleum, natural gas, and coal) are still going on, a very long time is needed for significant quantities of these fuels to be produced. We can expect that the amount of fossil fuels that will be formed in the next several thousand years will be negligible. The existing fossil-fuel supply can therefore be regarded as a nonrenewable resource.

Estimates of the fossil-fuel supplies remaining in the world are shown in Fig. 21.7. The entire circle represents the original source, the white portion shows how much has been used up, and the gray portion indicates the best estimate of how much remains. Although it may appear from

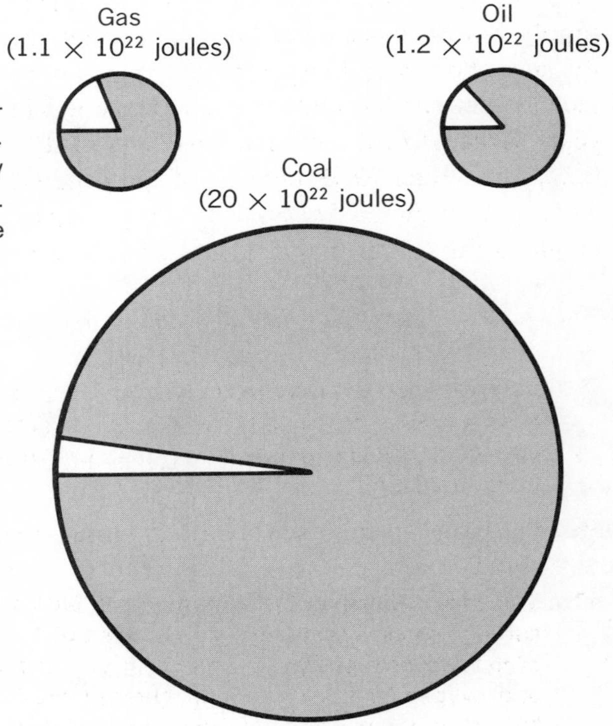

Gas
(1.1×10^{22} joules)

Oil
(1.2×10^{22} joules)

Coal
(20×10^{22} joules)

Fig. 21.7 Estimated remaining resources of fossil fuels. In each case the energy figure is the original total. The dark areas show the percentages still unused.

this diagram that we have only dented the original supply, increased worldwide energy demands have increased rates of consumption so greatly that we shall rapidly exhaust it. Some estimates indicate that the world coal supply will last another two or three hundred years, whereas the petroleum supply can serve as a major source of energy for only another 70 or 80 years. Nuclear fuel supplies, however, are nearly unlimited. The use of nuclear-energy plants has just begun. With uranium as the primary fuel (and the possibility of perfecting thorium reactors), we have a potential source of energy millions of times larger than all the reserves of fossil fuels.

Although man has made many useful technological advances to utilize energy, they have had a profound effect on the natural environment. Dams that back up water affect the ecological systems in the low lands and river beds. The combustion of fossil fuels by factories, power plants, automobiles, and aircraft adds foreign particles, carbon dioxide, and other undesirable substances to the atmosphere. We do not yet know what the overall global effects will be. It is clear, however, that as we increase our use of nature's energy sources, we must know what side effects we are causing.

Without the sun as a source of radiant energy, life as we know it today would cease to exist on the earth. It is natural to ask how long this supply can be expected to last. As the end result of nuclear reactions on the sun, about 4×10^{26} joules of energy are radiated out into space each second! The sun has been transferring energy at this rate for several billion years. Because of the tremendous mass of the sun, however, after all this time, more than 99.99 percent of the original fuel supply still remains! At least for billions of years to come, we apparently have nothing to worry about as far as our major energy source is concerned.

For Home, Desk, and Lab

8 How many gallons of gasoline would have to be burned to produce enough thermal energy to equal the energy of an average hurricane? (The heat of reaction of gasoline is about 5×10^7 joules/kg, and 1 gallon of gasoline has a mass of 2.5 kg.)

9 Can you think of ways to harness the energy from the incoming and outgoing tides?

10 The sun provides on the average about 0.01 calorie/cm^2/sec at the earth's surface. Assume that a leaf with an area of 10 cm^2 utilizes about 10 percent of the available energy in converting CO_2 and water into glucose ($C_6H_{12}O_6$) and oxygen by the process of photosynthesis. In this process about 4×10^3 cal of energy (from sunlight) is required to produce 1 g of glucose.

For Home, Desk, and Lab

About how many days will it take the leaf described above to produce 1 g of glucose? (Assume 12 hours of sunlight per day.)

11 The daily food intake for a mountain climber must provide about 5×10^6 calories. Sugar provides about 4×10^3 cal per gram.
   a) How much sugar per day would provide this energy?
   b) What is the increase in gravitational potential energy when a climber climbs a mountain 2,000 meters high?

12 A 1,000 kg car travels 15 km in going up a mountain 1 km high. In the process the car uses 3 kg of gasoline.
   a) What is the gain in the car's gravitational potential energy?
   b) How much energy is released by the burning gasoline? (The heat of reaction of gasoline is about 5×10^7 joules/kg.)
   c) What is the car's efficiency in climbing the hill?

13 From your data in Expt. 16.4, A Motor Lifting an Object, calculate the efficiency of the motor you used.

14 How much water would be heated 2°C by the heat released into Salem Harbor each hour by the electric generating plant there?

15 The electric wall outlet in your house is a supplier of energy. Trace this supply step by step as far back as you can.

16 What is the source of the energy supplied by a battery that runs an electric motor?

Answers to Problems Marked with a Dagger (†)

Chapter 11

1. (a) Indirectly
   (b) Directly
   (c) Indirectly
   (d) Indirectly
   (e) Indirectly
4. (a) Yes
   (b) 1:1
5. The student inadvertently placed one tube over the positive electrode and collected oxygen in this test tube.
7. The charge flowing past point A equals that flowing past point B and is one-half as much as that flowing past point C.
9. 10 cm^3
12. The charges are equal.
13. To be useful, a measuring device must have a minimal effect on the system that it is measuring. If the thermometer is too large, then bringing it in contact with the object whose temperature it is to measure would cause an appreciable change in the temperature of the body. By making thermometer bulbs much smaller than the object whose temperature is being measured, this effect is made negligible.
15. 20 amp sec
16. 27 sec
18. The needle would gradually move to the right.
20. About 20 cm^3

Chapter 12

1. The tube in which hydrogen collects.
5. There is no net change of mass at either nickel electrode nor in the solution. When water is electrolyzed, the mass of the electrodes is unchanged. The gases set free came from the water, and are not restored to the water when the battery terminals are reversed.
7. (a) 2
   (b) 9.6×10^{-3} g
11. (a) MgO
    (b) CrO$_3$
    (c) Al$_2$O$_3$
12. One elementary charge
14. So little charge would pass through the solution that no change in mass could be measured.
16. HgCl, HgCl$_2$

Chapter 13

6. The zinc would become coated with copper from the copper sulfate solution. The coating is very loose and will wash off easily. However, if all

177

Answers to Problems

the zinc is eventually "eaten" away, the steel becomes plated with copper.
8. Connect an ammeter and battery in series between the two remaining connections. Connect the heater circuit. If the ammeter shows a current, then the positive end of the battery is connected to the plate terminal.
11. About 7×10^6 sec $\approx$ 80 days
12. No.
13. (a) Left to right in both cells
    (b) Left to right
    (c) Counterclockwise

Chapter 14

1. 1.5×10^4 cal
4. 40 cal/°C
5. 2,000 cal/°C
7. (a) 66 cal/°C
   (b) 0.033 (cal/g)/°C
9. The cold water gained 4,000 cal. The hot water lost 4,000 cal.

Chapter 15

1. (a) 4.30 cal/°C
   (b) 17.2 cal
3. 30.95°C
4. There will be considerable heat loss throughout the run. This will yield too small a temperature change.
5. (a) 69 cells
   (b) 102 cells
6. For a 6.0-volt battery four 1.5-volt cells must be connected in series. A 90-volt battery must contain 60 cells.
11. (a) 4 cal/amp-sec
    (b) 17 joules/amp-sec
    (c) 17 volts
12. 8 joules
13. 0.52 amp
14. 1.08×10^6 joules
15. 900 joules
16. (a) 2 volts
    (b) 4 ohms

Chapter 16

1. 108 joules
6. Their temperature rise will be the same.
7. The object-lifting motor will not get as hot as the free-running motor.

Chapter 17

1. They are equal.
2. The changes are equal but opposite.
8. The change in gravitational potential energy will increase four times.
9. The 6-kg object.
10. 1.1°C
11. 3.6°C

Chapter 18

2. 95 cal/g
3. This would result in an apparent "missing" heat that would be too large.

Chapter 19

3. Probably not quite as high because of increased bearing friction and increased air resistance.
4. 0.1 m/sec
5. Halving the speed gives 0.7°C
   Doubling the speed gives 11.2°C
6. (a) 1
   (b) $\sqrt{2} = 1.41$
9. (a) 4.5×10^5 joule
   (b) 45 m
11. 3/1
14. The average speed of bromine molecules should be less than that of molecules of air when both gases are at the same temperature.

Chapter 20

2. A definition

Index

A

Ammeter, 12–14
Ampere (unit), 14
Anode, 41
Atomic masses:
    and charge per atom, 26–27
    table of, 21
Atomic model and vaporization, 117
Atomic potential energy, 115–120
Atomic vibrations, 162–163
Atoms, charged, 46–48
Automobile, efficiency of, 170–172

B

Battery cells. *See also* Electrolysis
    in circuits, 48–50
    and corrosion, 37–39
    Daniell cells, 32–34, 76–79
    direction of current in, 51
    dry cells, 34–36
    energy changes in, 151
    heat produced in, 69–70, 76–79
    plating cells, 20–22, 27–29, 48–49
Boiling points (table), 115
Boyle's law, 150

C

Calorie (unit), 56–58
Calorimeter, 54
Cathode, 41
Cells, *see* Battery cells
Charge, *see* Electric charge
Charge meter, 3–4
    effect of, on electric circuit, 12–14
Charge per atom of elements (table), 24, 29
Chemical energy, 120–121
    in the photographic process, 158
    from photosynthesis, 168–170

Circuit, *see* Electric circuit
Collisions of gas molecules, 137
Condensation, 112–114
    and hurricanes, 166
Conductors, electrical, 75
Conservation of charge, 10–11
    in a circuit, 49
Conservation of energy, 159–160, 161–163
Conservation of masses, 150
Constant proportions, 25–27, 29–30
Cooling, 61
Corrosion, 37–39

D

Daniell cell, 32–34, 76–79
Decomposition of water, *see* Electrolysis
Definitions and laws, 150–151
Density
    formula for, 150
    of gases, 141
    of a vapor, 115
"Disc gas" machine, 136–140
    with few molecules, 161–162
Dry cells, 34–36

E

Efficiency, 170–173
Effusion of gases, 140–146
Elastic potential energy, 105–106, 163
Electric charge
    of atoms, 19–20, 24
    conservation of, 10–11
    defined, 1
    and electric current, 14–17
    of electrons, 45–46
    and electroplating, 20–22
    flow of, in a circuit, 7

flow of, in a light bulb, 4–6
and heat, 65–67
measured with an ammeter and a clock, 16–17
motion of, in a circuit, 48–50
motion of, in a vacuum, 41–44
quantity of, 2–4
unit of, 25
Electric circuit, 3
effect of charge meter on, 12–14
motion of charge in, 48–50
series and parallel, 7
Electric current, 14–17
direction of, 49–51
and electric power, 72
and heat, 65–67
Ohm's law for, 74–75
Electric motor
free running, 90, 151
heat capacity of, 86–89
heat produced in, 84–86
lifting an object, 91–92, 151
Electric power, 72
Electric power plant, 172–173
Electrical conductors, 75
Electrical resistance, 73
Electrical work, 71–72, 90–91
Electrolysis. *See also* Plating cells
and charge per atom, 19–20
and elementary charge, 24
and flow of charge, 1–4
heat produced in, 117–120
Electrons, 45–46
Electroplating, 20–22, 27–29
Elementary charge, 23–27
definition, 25
Elements
electric charge of, 24
masses of (table), 21
Energy
changes of, 151–156
chemical, 120, 158, 168–170
conservation of, 159–160, 161–163
defined, 94–95
kinetic, *see* Kinetic energy
of light, 156–158
nuclear, 160

potential, 95, 105–106, 115–120. *See also* Gravitational potential energy
of radiation, 156–158, 165, 168–169, 175
thermal, 95, 135, 153, 163
Energy converter, large-scale, 172–73
Energy losses, *see* Efficiency; Energy transfer
Energy sources, 174–175
Energy transfer, 151–156. *See also* Conservation of energy
and efficiency, 170–172
in a hurricane, 166
in photosynthesis, 168–170
in a power plant, 172–173
from the sun, 165, 168–169, 175
Evaporation, 110–112, 114–115
and formation of hurricanes, 166

F

Faraday, Michael, 25
Flashlight cells, 34–36
Formulas of compounds, 26–27, 30
Fuels, supply of, 174–175

G

Galvanized iron, 39
Gases
collisions in, 137
densities of, 141
and "disc gas" machine, 136–140
effusion of, 140–146
molecular masses of, 141
molecular speeds in, 137, 140–146
thermal energy of, 135
Gravitational potential energy, 94
changes in, 153
and elastic potential energy, 106
and height, 100
and kinetic energy, 125–127, 129–132
and mass, 95–99
and path traversed, 100–103
and thermal energy, 94–95

H

Heat
 of condensation, 112–114, 166
 and decomposition of water, 117–120
 and electrical work, 71–72
 and electric motor, 90–92
 and electrolysis, 117–120
 and evaporation, 110–112, 114, 166
 generated in a contracting spring, 104
 and loss of speed, 123–125
 lost in cooling, 61
 quantity of, 54, 58
 of reaction (table), 121
 per unit charge, 68–70, 79
 of vaporization, 110–112, 114–115, 166
Heat capacity, 59–60, 62, 151
 of an electric motor, 86–87
Heater, electric, 65–71
 and resistance, 73–75
 in a vacuum tube, 41–43
 and work, 71–72
Helmholtz, Hermann von, 25, 159
Hurricane, energy in, 166
Hydroelectric plants, 173
Hydrogen, electric charge of, 19–20
Hydrogen cell, 3–4
 effect of, on electric circuit, 12–14

I

Inclined plane, 100–103, 163
Ions, 46–48

J

Joule (unit), 71
Joule, James, 95, 159–160

K

Kilowatt hour (unit), 72
Kinetic energy
 changes in, 153
 defined, 125
 of gases, 135
 and gravitational potential energy, 125–127, 129–132
 and mass, 128, 130, 133–134
 molecular, 135
 and speed, 128–131
 and temperature, 135, 140

L

Laws and definitions, 150–151
Light, energy of, 156–158
Light bulbs
 flow of electric charge in, 4–6
 energy associated with, 156

M

Mass
 and gravitational potential energy, 95–99
 and kinetic energy, 128, 130, 133–134
Masses
 atomic, 21, 26–27
 of gas molecules, 141
Molecular speed in gases, 137, 140–146
Motor, electric, *see* Electric motor

N

Nonreversible processes, 161–163
Nuclear energy, 160
Nuclear-energy plants, 175

O

Ohm (unit), 74
Ohm's law, 74–75
Open-circuit voltage, 79

P

Parallel circuit, 7

Photographic process, 158
Photosynthesis, 168–170
Plating cells, 20–22, 27–29, 48–49
Potential energy, 95
    atomic, 115–120
    elastic, 105–106
    gravitational, see Gravitational potential energy
Power, electric, 72

R

Radiant energy, 156–158
    from the sun, 165, 168–169, 175
Radiation absorbed in the atmosphere, 165
Radiometer, 158–159
Recharging cells, 36
Resistance, electrical, 73–75
Resistors, 75
Rust, 37–39, 120

S

Series circuit, 7
Solar energy, 165, 168–169, 175
Specific heat, 59–60, 62
Springs, 104–106, 155–156
Steam engine, 159
Storms, energy of, 166

T

Temperature
    effect of, on electrical resistance, 75
    and kinetic energy, 135, 140
Thermal energy, 95
    changes in, 153
    and conservation of energy, 163
    of a gas, 135
Tides, 163
Turbines, 173

V

Vacuum tube, 41–44
Vaporization, 110–115, 166
Volt (unit), 69
Voltage, 69–70
    open-circuit, 79
Voltmeter, 69

W

Water, decomposition of, see Electrolysis
Watt (unit), 72
Watt, James, 72
Work, electrical, 71–72, 90–91